Varieties of Continua

Varieties of Continua

From Regions to Points and Back

Geoffrey Hellman and
Stewart Shapiro

OXFORD
UNIVERSITY PRESS

OXFORD
UNIVERSITY PRESS

Great Clarendon Street, Oxford, OX2 6DP,
United Kingdom

Oxford University Press is a department of the University of Oxford.
It furthers the University's objective of excellence in research, scholarship,
and education by publishing worldwide. Oxford is a registered trade mark of
Oxford University Press in the UK and in certain other countries

First Edition published in 2018

Impression: 1

Published in the United States of America by Oxford University Press
198 Madison Avenue, New York, NY 10016, United States of America

British Library Cataloguing in Publication Data
Data available

Library of Congress Control Number: 2017950767

ISBN 978–0–19–871274–9

Printed and bound by
CPI Group (UK) Ltd, Croydon, CR0 4YY

Contents

Preface

This work began several years ago with a series of discussions on competing conceptions of continua in mathematics, informed by late twentieth-century developments, such as Robinsonian non-standard analysis and the more radical, non-classical smooth infinitesimal analysis, that rehabilitated earlier notions of infinitesimals. At first our plans were rather modest—to do a comparative study of leading approaches to the continuous, assessing the trade-offs involved, in a more systematic and comprehensive way than what was already available in the literature. In particular, we were struck by the historical transition from the long-standing, Aristotelian non-punctiform conception to modern punctiform analysis based on the method of limits due to Bolzano, Cauchy, Weierstrass, Cantor, Dedekind, et al., which rapidly became the dominant mainstream approach. Along with the contrast between non-punctiform, or regions-based, conceptions and the modern view of continua as entirely composed of points, there is also the contrast between potential infinity—again going back to Aristotle—and the recognition of actual infinities in mainstream modern mathematics. Here we were struck by the circumstance that restrictions to intuitionistic logic, also associated with potential infinity, have been motivated in radically different ways, on the one hand by demands of constructivity (by Brouwer, Bishop, et al.), and, on the other, by restrictions to "smooth manifolds" (as in smooth infinitesimal analysis (SIA), especially as described by John Bell), although this latter way of motivating the restriction has been challenged by one of us. Also, we were motivated to assess intermediate, semi-constructive systems of predicative analysis (going back to Russell, Poincaré, Weyl, and perfected in work of Feferman et al.), based on classical logic, recognizing actual infinity but only of predicatively justifiable countable totalities.

Given this variety of systems and motivating ideas, it seemed clear to us that "continuity" is a cluster-concept, realizable in multiple, often competing ways, and that no single system could do justice to all the desiderata emerging from mathematical and scientific practice, past and present.

It then occurred to us that it ought to be possible to avoid the shift to intuitionistic logic, along with the apparatus of nil-potent infinitesimals, while still basing analysis and geometry on the idea of regions or "chunks" of a given space or manifold, eschewing points, at least as parts of regions. Could not one remain "classical" in one's proof theory and treatment of the infinite while still realizing a thoroughgoing non-punctiform conception of continua? We were, of course, already familiar with ideas of Whitehead and others on recovering "points" as defined by suitably nested regions of a space. We were, however, surprised at how little had been done in working out such very natural ideas rigorously, especially by deploying the resources of atomless mereology, which seems well-suited for such a project. In particular, Tarski had begun such a project in early work with Leśniewski, supplementing mereology with the sole primitive "sphere" in order to recover classical, three-dimensional geometry. But on closer examination we realized that Tarski's ingenious definitions were not supplied with autonomous axioms, beyond those of mereology, on the mathematical primitive, "sphere", framed directly in terms of regions, so that the project remained in an incomplete state.

So we set ourselves the task of completing such work, but beginning with the simple case of a continuum of one dimension and then trying to extend such a treatment to higher-dimensional spaces. This in turn involved us in seeking to recover, on the same basis of regions, metrical structures of Euclidean and then non-Euclidean geometries, ultimately resulting in the core positive developments of the present work.

It was clear to us from the outset that there is a strong historical dimension to our project, rooted as it is in ideas going back to Aristotle. Indeed, we call our system for a one-dimensional regions-based ("gunky") continuum "semi-Aristotelian", since it adheres strictly to Aristotle's desideratum of not being constituted of points while at the same time making free use of actual infinities, both geometric and set-theoretic (equivalently described via logic of plurals). This in turn naturally led us to ask how we might formulate more fully Aristotelian theories of continua, enforcing restriction to potential infinities only. As Øystein Linnebo had already worked out formalism based on modal logic for expressing potential infinity, we pursued a three-way collaboration with him for a chapter devoted to this topic.

At this stage, then, we had developed systems of analysis and geometry that are "semi-Aristotelian", non-punctiform but "classical" with respect to actual infinities, and we had "more Aristotelian" systems that were both non-punctiform and confined to the potentially infinite. That left the further possibility of predicative systems, lying between the more Aristotelian and our semi-Aristotelian ones, based on regions, using classical logic (without modality), but admitting only countably infinite totalities (corresponding to well-founded set theory without the power–sets axiom). The development of such predicative systems then rounds out our formal work presented here.

We then turned our attention to discussing some of the interesting philosophical implications of the formal developments, including the systematic comparisons of the different approaches that motivated our project in the first place, but also implications for various discussions of contemporary analytic metaphysics pertaining to applications of geometry and analysis to material objects of the natural world, such as how to understand or model "contact" of material objects, whether to admit point-like objects, and so forth. As explained below, many of these discussions err in simply assuming that the geometry being applied pertains to punctiform spaces, whereas various conundra evaporate in the context of "gunky" spaces.

Finally, to round out this volume, we return to our original theme of comparisons and trade-offs among the diverse conceptions of continua, now including the regions-based systems constructed here.

We incurred many debts along the way. In the initial development of the one-dimensional, semi-Aristotelian system, we were helped by Roy Cook, a former student of one of us and a current colleague of the other, and by graduate students at our home institutions, notably Shay Logan at Minnesota and Patrick Reeder at Ohio State. Either together or separately, we gave presentations of early versions of some of this material at a variety of conferences, workshops, and colloquia, and were enlightened by comments from the audiences. The list includes the University of Minnesota, the Ohio State University, several meetings of the Midwest Philosophy of Mathematics Workshop, held annually at Notre Dame (and once in Illinois), the Foundations Interest Group at the Minnesota Center for Philosophy of Science, Logica, the Philosophy of Science Association, the University of Sheffield, the University of Oslo, the University of California at Davis, the Hebrew University of Jerusalem, the University of Connecticut, the University of Southern California, and the University of St Andrews. Thanks to anonymous reviewers from Oxford University Press, *Philosophia Mathematica*, *Logic and Logical Philosophy*, *Iyyun*, and the *Review of Symbolic Logic*. Thanks also to Frank Arntzenius, Jeremy Avigad, John Bell, Einar Duenger Bohn, Sarah Broadie, John Burgess, Ben Caplan, Tim Carlson, Aaron Cotnoir, Philip Ehrlich, Thomas Forster, James Franklin, Harvey Friedman, Giangiacomo Gerla, Charles Geyer, Rafał Gruszczyński, Jeremy Heis, Eli Hirsch, Christopher Hookway, Jesse Houchens, Øystein Linnebo, Charles David McCarty, Marco Panza, Andrzej Pietruszczak, Carl Posy, Agustín Rayo, Barbara Sattler, Wilfried Sieg, Allan Silverman, Reed Solomon, Gabriel Uzquiano, Achille Varzi, and Keren Wilson. Sincere apologies to all of those who should be mentioned here, but are not. We owe a large debt of gratitude to Haidar Al-Dhalimy for a most careful proofreading of earlier drafts, developing the LaTeX code for our figures, and for help with the final proofreading and the index. Finally, thanks to Peter Momtchiloff, of Oxford University Press, and his staff, for encouraging this project and for guiding it through the various stages of publication.

1

The Old Orthodoxy (Aristotle) vs the New Orthodoxy (Dedekind–Cantor)

1.1 Common Ground

In some ways, Aristotle came remarkably close to the contemporary, "classical", or "orthodox" account of continuity, especially in comparison with his main opponents on these matters, the atomists (see White [127] and Miller [87]).[1] First, for Aristotle, a continuum is *infinitely divisible*, in the sense that it has no smallest parts. In the case of one dimension, the idea is that any line segment can be bisected (or divided into any finite number of parts). The results of a bisection are two segments, each of which can be bisected, and so on indefinitely. The atomists, or at least those radical atomists who applied their view to space itself and time itself, insist that there are smallest regions, those that cannot be further divided. Zeno's paradox of the Stadium seems to be aimed at those theorists. For Dedekind–Cantor, of course, a continuum has no smallest *extended* parts, although one can think of a point as an unextended part of a line or line segment.

In at least one place, Aristotle seems to say that infinite divisibility is *sufficient* for continuity:

> . . . no continuous thing is divisible into things without parts. Nor can there be anything of any other kind between; for it would be either indivisible or divisible, and if it is divisible, divisible either into indivisibles or into divisibles that are always divisible, in which case it is continuous.
>
> (*Physics* VI, 231b11–15)

Second, Aristotle's continuum is Archimedean. This has two formulations, the second a sort of inverse to the first:

(i) Let a and b be any two magnitudes, of the same dimension. Then there is a natural number m, such that the result of adding a to itself m-times is larger

[1] It is ironic that the Dedekind–Cantor account is called "classical" since, in the present context, we *contrast* that account with Aristotle's.

than b. As Aristotle puts it (*Physics*, 8, 10): "By continually adding to a finite quantity, I shall exceed any definite quantity".

In Euclid's *Elements* (Book V, Definition 4), this principle appears as a definition: "Magnitudes are said to have a ratio to one another which are capable, when multiplied, of exceeding one another". The assumption is that any two magnitudes of the same dimension "have a ratio" to each other.

(ii) Let a and b be any two magnitudes. Suppose we remove at least half of a, and then remove at least half of the remainder, etc. Continuing in this way, we will eventually produce a magnitude smaller than b. As Euclid (Book 10, Proposition 1) puts it:[2]

> Two unequal magnitudes being set out, if from the greater there be subtracted a magnitude greater than its half, and from that which is left a magnitude greater than its half, and if this process is repeated continually, there will be left some magnitude less than the lesser magnitude set out.

The radical atomists presumably reject the second principle, although some of them may accept the first version.

1.2 Points

For present purposes, of course, the differences between the Aristotelian and the contemporary, Dedekind–Cantor account of continuity are more salient. Almost all of these differences are tied, in one way or another, to Aristotle's rejection of the actual infinite, although we are loath to speculate on the conceptual and explanatory connections between the various features. We begin with the matter of points.

To put things a bit anachronistically, Aristotle argues that time, space, and motion all have the same structure. The reason seems to be that motion is to be understood as movement in space through time. It follows that, for Aristotle, if any one of time, space, or motion is punctiform—composed entirely of points, or point-like things—then so are the others. And if any one of time, space, or motion is composed of extended atoms, then so are the others, etc. It seems that everyone in the ancient, medieval, and early modern debates accept that much.

As we just saw, Aristotle rejects the atomistic hypothesis that there are extended atoms in space or time. For present purposes, the central feature of the Aristotelian account—and one primary difference with the received Dedekind–Cantor one—is that continua are *not* composed of points. In other words, for Aristotle, continua are not punctiform. Sticking, here, to one dimension, the parts of a line segment are other line segments (or sums thereof).

[2] Euclid proves this from the first Archimedean principle (Book 5, Definition 4), thus presupposing that the given magnitudes "have a ratio" to each other. He also invokes some properties of unequal magnitudes. The method of exhaustion, traced to Eudoxus, relies on this principle.

Michael J. White [127], p. 29 attributes the following "principle of non-supervenience of continuity" to Aristotle:

N-SC. Each partition of a continuous magnitude into proper parts yields parts each of which is pairwise continuous with at least one other part.

We will soon get to what Aristotle means for two things to be "continuous with each other". For now, the crucial feature is that if two things are continuous, then nothing of the same kind can come between them. For Aristotle, as for Euclid, as in contemporary Euclidean geometry, between any two points, there is a line segment. As Aristotle puts it, "that which is intermediate between points is always a line, and that which is intermediate between moments is a period of time" (*Physics*, Book 6, 231a1). So no two points or instants can be continuous (or contiguous) with each other. Thus, the principle N-SC entails that a line segment, or an interval of time, does not have points, or instants, as parts.[3]

To repeat, and to labor the obvious, this is one key place where the contemporary Dedekind–Cantor conception of the continuum parts company with Aristotle. The now "orthodox" account does construe a line segment as (nothing more than) a set of points, each of zero length. And a time interval is a set of instants, each of zero duration. The Dedekind–Cantor theory is sometimes called a "point–set" account.

To be sure, Aristotle does recognize points, in some sense. Indeed, points play a crucial role in his views on continuity and indecomposability. For Aristotle, points are the limits, or endpoints, of line segments.[4] The points interior to a line segment exist only potentially, not actually. An interior point represents a place where a segment *could be* divided. So to bisect a line segment is to indicate a place where it could be broken. But unless the segment *is* divided there, the point does not exist—it is not actual.

This, we take it, is the main feature of Aristotle's response to Zeno's paradox of Division. Suppose that someone, Ms. Walker, wishes to cover a distance of, say, 100 meters. She must first walk the first 50 meters. Call that Walk 1. Then she must walk the next 25 meters. Walk 2. Then the next 12.5 meters. Walk 3. And on it goes. So in order to reach her destination, Ms. Walker must complete an infinite number of Walks. Zeno concludes that she cannot do so, and, thus, that motion is impossible.

Aristotle, of course, follows common sense, and disagrees with the conclusion. Motion is indeed possible (presumably because it is actual). Aristotle's views on points take on much of the burden of explaining what goes wrong with Zeno's argument.

[3] White's principle of non-supervenience presupposes that continuous magnitudes (such as line segments) *can* be partitioned into proper parts. This runs against a long-standing theme that a continuous magnitude *cannot* be divided, cleanly, into proper parts. As we shall soon see, Aristotle himself held a version of this "indecomposability". According to contemporary intuitionism, continua are indecomposable in a rather strong sense. See Chapter 8.

[4] Also, for Aristotle, lines are just the edges of plane figures, and plane figures are just the faces of three-dimensional physical objects.

Consider the endpoint of Walk 1, a point 50 meters from the start of the journey. For Aristotle, that represents a place where Ms. Walker *can pause* on her journey, either temporarily or permanently. There are infinitely many places where she could stop. But unless she *does* stop there (or someplace else), there simply is no separate Walk 1 to consider. There is just a single event, her walk of the entire 100 meters. That walk is not the sum of Walk 1, Walk 2, . . .

We do not claim that this resolves every version of the paradox. Aristotle does not consider a so-called "staccato" scenario like the following: Ms. Walker takes 30 minutes to complete Walk 1, and then pauses for 15 minutes. Then she takes 7.5 minutes for Walk 2, and pauses for 3.75 minutes. Then she takes 1.875 minutes for Walk 3 and pauses for half of that, etc. It still takes 60 minutes to cover the full 100 meters and, by Aristotle's own lights, Ms. Walker *does* have to complete an infinite number of walks first.[5]

To summarize, it seems that, for Aristotle, line segments have actual endpoints—but such endpoints are decidedly not *part* of the line segment. They are just its boundaries. Each endpoint is metaphysically tied to the segment it bounds, and cannot be considered in isolation from the segment, in the same sense as the smile of the Cheshire cat cannot be considered apart from the cat. The interior points on a line segment exist only potentially. There is a potential infinity of such interior points, but not an actual infinity of them (see Lear [70]).

In *Physics* 8.8 (263bg3–6), we find:

> So that if someone asks whether you can traverse an infinity either in time or in length, we must say that in a way you can, and in a way you cannot. For you cannot traverse an infinity of actually existing [divisions] but you can of potentially existing ones.
>
> (quoted in Sorabji [116], 213)

We will turn to matters of infinity soon.

1.3 Succession, Contiguity, and Continuity

In the *Categories*, Aristotle distinguishes discrete from continuous entities:

> Discrete are number and language; continuous are lines, surfaces, bodies, and also, besides these, time and place. For the parts of a number have no common boundary at which they join together . . . Nor could you ever in the case of number find a common boundary of its parts, but they are always separate. Hence number is one of the discrete quantities. Similarly, language also is one of the discrete quantities . . . For its parts do not join together at any common boundary. For there is no common boundary at which the syllables join together, but each is separate in itself. A line, on the other hand, is a continuous quantity. For it is possible to find a common boundary at which its parts join together, a point. (4b22)

[5] It is unclear whether this "staccato" scenario would be at all relevant to the debate at the time. Zeno was arguing that motion (as an instance of change) is impossible. No one in the ancient debate would regard the "staccato" walk as possible. Thanks to Sarah Broadie here. See [127], 53–72 and [70].

Number, here, means "positive natural number". For Aristotle, natural numbers are always *numbers of* some things, treated as discrete. And "line" is what we would call a line segment. For Aristotle, what makes line segments "continuous" (with each other) is that when they join together, there is a common boundary, namely a point.

In Book 5 of the *Physics*, Aristotle introduces a common-sense notion:

> A thing is in succession when it is after the beginning in position or in form . . . and when further there is nothing of the same kind as itself between it and that to which it is in succession, e.g. a line or lines, if it is a line, a unit or units if it is a unit, a house if it is a house. (226b34)

To follow one of his examples, think of the houses (after the first) in a row on a given street. They are in succession because there are no houses between any consecutive two.

Aristotle defines two things to be *contiguous*, or *in contact*, if they are in succession, in such a way that nothing of the same kind *can go between*: "Things are said to be in contact when their extremities are together" (226b21). Think of a pair of adjacent books on a tightly packed shelf. Those are (or appear to be) in succession and contiguous (i.e. in contact).

He goes on to define another notion:

> The continuous is just what is contiguous, but I say that a thing is continuous when the extremities of each at which they are in contact become one and the same and are (as the name implies) contained in each other. Continuity is impossible if these extremities are two. This definition makes it plain that continuity belongs to things that naturally, in virtue of their mutual contact form a unity. (227a6)

So the aforementioned books on the shelf are not continuous, since no matter how close to each other they get, each retains its own boundaries. When contiguous, they remain *two books*, each with its own boundary. In contrast, when continuous things, such as line segments, are brought into contact, the common boundary between them is absorbed, and they become a single line, a "unity". For a more physical example, think of two bodies of (liquid) water, or consider what happens when one leaves two loaves of wet dough to rise before being baked into bread. If the loaves are placed too close together, when they come into contact with each other, they will absorb each other, and merge into a single loaf—with no discernible boundary between them.

It is curious that Aristotle defines continuity as a *relation* between things: they are continuous (or not) *with each other*. Yet he also speaks of continuous *things*, suggesting that continuity is a (monadic) property. Perhaps the idea is that a substance is continuous if whenever neighboring parts are considered, they are continuous with each other. Or perhaps continuous things only have parts potentially. Think, for example, of a limb of a tree. To break it up into parts would be to destroy it, or at least to destroy its unity.

In this last passage, Aristotle sounds a theme that plays a role in the subsequent development of continuity, up to, but not including, the advent of the Dedekind–Cantor account. A continuous thing, for Aristotle, forms a *unity*. There is something that binds it together, and makes it One thing. In a slogan, continuous substances are viscous.[6] The contemporary, Dedekind–Cantor continua are not like this. A line, for example, is just a set of points. And a set can be broken cleanly and arbitrarily (at least assuming classical logic) into its various subsets.

In Book 6 of *Physics*, Aristotle summarizes the definitions, and sounds a theme that motivates the present project:

> Now if the terms "continuous", "in contact", and "in succession" are understood as defined above—things being "continuous" if their extremities are one, "in contact" if their extremities are together", and "in succession" if there is nothing of their own kind intermediate between them—nothing that is continuous can be composed of indivisibles: e.g. a line cannot be composed of points, the line being continuous and the point indivisible. (231a1)

> ... since indivisibles have no parts, they must be in contact with one another as whole with whole. And if they are in contact with one another as whole with whole, they will not be continuous: for that which is continuous has distinct parts ... as we saw, no continuous thing is divisible into things without parts ... Moreover, it is plain that everything continuous is divisible into divisibles that are infinitely divisible. (231b)

In *Metaphysics* 3.5 (1002a28–b11), Aristotle connects this with his views on points:

> For as soon as bodies come into contact or are divided, the boundaries simultaneously become one if they touch and two if they are divided. Hence, when the bodies have been put together, one boundary does not exist, but has ceased to exist, and when they have been divided, the boundaries exist which did not exist before (for the point, being indivisible, was not divided into two). (quoted in [116], 11)

It is these views on continuity and boundaries—points in particular—that underlie the Aristotelian thesis of indecomposability. For Aristotle, the reason one cannot break a continuous object, such as a line segment, cleanly into pieces is that doing so *creates*, or makes actual, the boundaries of the resulting segments. Something new comes into being (or into actuality).

Other theories of continuity, such as intuitionistic analysis and smooth infinitesimal analysis show a much stronger sense of indecomposability (see, for example, Dummett [39] and Bell [15], and Chapter 8 for a brief summary). In those theories, it is not the case that an interval can be broken up into pieces, cleanly, with nothing left out. Any attempt to break up a segment *will leave something out*. The metaphor is that when you cut a line segment, something will stick to the knife.

1.4 Infinity

We now turn to what is probably the most far-reaching difference between Aristotelian continua and the contemporary Dedekind–Cantor account. Beginning with Aristotle, almost every major philosopher and mathematician before the nineteenth

[6] Thanks to Marco Panza, Sarah Broadie, Keren Wilson, Carl Posy, and Barbara Sattler here.

century rejected the notion of the actual infinite. They all argued that the only sensible notion is that of potential infinity—at least for scientific or, later, non-theological purposes. In *Physics* 3.6 (206a27–9), Aristotle wrote, "For generally the infinite is as follows: there is always another and another to be taken. And the thing taken will always be finite, but always different." As Richard Sorabji [116] (322–3) puts it, for Aristotle, "infinity is an extended finitude" (see also [70], [71], [81]).

Aristotle, along with ancient, medieval, and early modern mathematicians, recognized the existence of certain *procedures* that can be iterated indefinitely, without limit. Examples are the bisection and the extension of line segments. Ancient mathematicians made brilliant use of such procedures. For example, the method of exhaustion, a kind of forerunner to integration, was employed to calculate the areas of curved figures in terms of rectilinear ones.

What was rejected are what would be the end results of applying these procedures infinitely often: self-standing points, infinitely long regions, and infinite (or, in modern terms, transfinite) sets. This, of course, is connected to Aristotle's views on points. If, contrary to Aristotle (and contrary to the radical atomists), we are to think of a line segment as a collection of (actually existing) points, then that collection is indeed infinite—actually infinite. Aristotle, of course, rejects the very existence of actually infinite collections and the thesis that lines are composed of points.

In *On generation and corruption*, he wrote:

> For, since no point is contiguous to another point, magnitudes are divisible through and through in one sense, and yet not in another. When . . . it is admitted that a magnitude is divisible through and through, it is thought that there is a point not only anywhere, but also everywhere in it: hence it follows that the magnitude must be divided away into nothing. For there is a point everywhere within it, so that it consists either of contacts or of points. But it is only in one sense that the magnitude is divisible through and through, viz. in so far as there is one point anywhere within in and all its points are everywhere within it if you take them singly. (317a3–8)

Jonathan Lear [70] argues that it is not the existence of iterated procedures that makes for Aristotelian potential infinity. The matter concerns the *structure* of geometric magnitudes:

> . . . it is easy to be misled into thinking that, for Aristotle, a length is said to be potentially infinite because there could be a process of division that continued without end. Then it is natural to be confused as to why such a process would not also show the line to be actually infinite by division. . . . [I]t would be more accurate to say that, for Aristotle, it is because the length is potentially infinite that there could be such a process. More accurate, but still not true . . . Strictly speaking there could not be such a process, but the reason why there could not be is independent of the structure of the magnitude: however earnest a divider I may be, I am also mortal. . . . even at that sad moment when the process of division does terminate, there will remain divisions which could have been made. The length is potentially infinite not because of the existence of any process, but because of the structure of the magnitude. (p. 193)

Lear highlights the above theme that, for Aristotle, a line segment is potentially infinite because there are infinitely many places where it *can be* divided. So, no matter how many times one divides a line, there will still be some of the line left. Lear concludes that Aristotle's thesis is "that the structure of the magnitude is such that any division will have to be only a partial realization of its infinite divisibility: there will have to be possible divisions that remain unactualized" (p. 194).[7]

For Aristotle, lines themselves are also only of finite length. That is, Aristotelian lines are what are now called line segments. As noted, each such line has two endpoints. Euclid, too, seems to allow only line segments, at least for the most part. The first postulate is "To draw a straight line from any point to any point". This, of course, only produces segments. The second postulate is "To produce a finite straight line continuously in a straight line". The idea, it seems, is that a straight line can be extended as far as one wishes, but there is little or no need to countenance infinitely long lines. The language of potential extension is invoked explicitly in the (in)famous fifth postulate: "That, if a straight line falling on two straight lines make the interior angle on the same side less than two right angles, the two straight lines, *if produced indefinitely*, meet on that side on which the angles are less than the two right angles" (emphasis added, of course).

Given his overall philosophy of mathematics, Aristotle may have had a problem with even the *potential* infinity of line segments, and thus with Euclid's second and fifth postulates. As noted above, Aristotle held that mathematical entities are abstracted from physical entities. Numbers, for example, are numbers of objects, such as numbers of cows, construed as indivisible. Lines are the edges of physical objects, such as cubes. Aristotle also held that the entire universe is finite in extent. It is a sphere. So there cannot be any line longer than the diameter of that sphere (*Physics*, Book 3, 7: 207b27). See Knorr [68], 121–2, and Hintikka [62]. We make no attempt to recapture this aspect of Aristotle's views.

To summarize, for Aristotle, as for Euclid, lines are not actually infinite (i.e. not infinitely long). But for Euclid, at least, lines are potentially infinite, in the sense that any line can be extended.

As noted, when it comes to the infinite—or at least infinite collections—views like Aristotle's were standard throughout the medieval and early modern period, through most of the nineteenth century. The greatest mathematical minds insisted that only the potentially infinite makes sense. Leibniz, for example, wrote:

> It could . . . well be argued that, since among any ten terms there is a last number, which is also the greatest of those numbers, it follows that among all numbers there is a last number, which is also the greatest of all numbers. But I think that such a number implies a

[7] It is generally held that Aristotle's own views on modality are vexed. We make no attempt to sort that out.

contradiction... When it is said that there are infinitely many terms, it is not being said that there is some specific number of them, but that there are more than any specific number.
(letter to Bernoulli, [72], III 566, translated in [76], 76–7, 87)[8]

... we conclude... that there is no infinite multitude, from which it will follow that there is not an infinity of things, either. Or [rather] it must be said that an infinity of things is not one whole, or that there is no aggregate of them. ([73], 6.3, 503, translated in [76], 86)

Yet M. Descartes and his followers, in making the world out to be indefinite so that we cannot conceive of any end to it, have said that matter has no limits. They have some reason for replacing the term "infinite" by "indefinite", for there is never an infinite whole in the world, though there are always wholes greater than others ad infinitum. As I have shown elsewhere, the universe cannot be considered to be a whole. ([74], 151)

And Gauss [47]:

I protest against the use of infinite magnitude as something completed, which is never permissible in mathematics.

For Gauss and Leibniz, as for Aristotle, as for a host of others, the infinite just *is* the limitlessness of certain processes; no actual infinities exist. The only intelligible notion of infinity is that of potential infinity—the transcendence of any (finite) limit.[9]

For at least the cases of interest here—regions, natural numbers, and the like—Georg Cantor argued for the exact opposite of this, claiming that the potentially infinite is dubious, unless it is somehow backed by an actual infinity:

I cannot ascribe any being to the indefinite, the variable, the improper infinite in whatever form they appear, because they are nothing but either relational concepts or merely subjective representations or intuitions (imaginationes), but never adequate ideas. ([27], 205, n. 3)

... every potential infinite, if it is to be applicable in a rigorous mathematical way, presupposes an actual infinite. ([28], 410–11)

We think it safe to say that this Cantorian orientation is now dominant in the relevant intellectual communities, especially concerning the mathematical domains mentioned above, with various constructivists as notable exceptions (stay tuned).

1.5 Plan

As noted above, the now orthodox Dedekind–Cantor account of continuity makes two major departures from the Aristotelian tradition. It takes continua to be composed of points, and it makes (heavy) use of actual infinity, à la Cantor. One of the

[8] The "contradiction" mentioned here might be the so-called "Galileo paradox", that with infinite collections, a proper subset can be equinumerous with a set. This, of course, is now a standard *feature* of infinite sets; a feature and not a bug.

[9] To be sure, Leibniz, Gauss, and a host of others were not completetely consistent on this. They were, after all, pioneers in the emergence of modern mathematics, not to mention their use of infinitesimals. A reader reminds us that while Leibniz rejected infinite *number*, he was, at times, a champion of actually infinite totalities. The exegetical issues are (well) beyond the scope of this work.

main conclusions of this study is that, of these two, the emergence of infinity is far more significant. The matter of whether continua are composed of points is, at most, a convenience.

Chapter 2 develops a "semi-Aristotelian" account of a one-dimensional continuum. Unlike Aristotle, it makes significant use of actual infinity, in line with current practice, of course. Like Aristotle, however, this account does not recognize points, at least not as parts of regions in the space. The formal background is classical mereology together with either (a weak) set theory, higher-order logic (understood with "standard semantics"), or a logic of plural quantification. The first-order variables range over regions of the space, and there is an axiom that entails that every region has a proper part. This, of course, is inconsistent with the existence of points (at least with points as parts of regions, as they are in ordinary, Dedekind–Cantor spaces).

In contrast to intuitionistic analysis, smooth infinitesimal analysis, and Errett Bishop's [19] constructivism, we show that it is possible to partition our "gunky line" into mutually exclusive and exhaustive discrete parts (e.g. to cleanly bisect intervals). This demonstrates the independence of "viscosity" and "indecomposability" from a nonpunctiform conception of the continuum.

It is surprising that such simple axioms as ours imply the Archimedean property and an interval analogue of Dedekind completeness (least-upper-bound principle). This is without an "extremal axiom" (to the effect that "these are the only ways of generating regions"), a principle that, in effect, just imposes completeness. We also establish an isomorphism with the Dedekind–Cantor structure of the real line $\mathbb{R}$ as a complete, separable, ordered field, and show how to define "points" as a sort of Whiteheadian "extensive abstraction". A "point", so construed, is an equivalence class of certain sets of regions. The "points" thus defined are isomorphic to the standard real line.

We also present some simple topological models of our system, establishing consistency relative to classical analysis. Indeed, our gunky theory and the more standard Dedekind–Cantor one are mutually interpretable and, indeed, definitionally equivalent. So, at least for a one-dimensional continuum, the existence of points, as parts of regions, is not necessary.

To be sure, ours is not the first attempt at a rigorous point-free (or pointless), regions-based account of the continuous, although we think it has some interesting features. We provide comparisons with some earlier accounts (although other comparisons are postponed until after we develop higher-dimensional versions of the theory).

Chapter 3, coauthored with Øystein Linnebo, presents a more Aristotelian theory, one that eschews both the existence of points (as parts of regions) and the use of actual infinity. Infinity enters the semi-Aristotelian theory, of Chapter 2, with our mereological principle of unrestricted fusion. In developing that theory, we sometimes take the fusion of an explicitly infinite set of regions. This occurs, for example, in the proof of the Archimedean property and in the recovery of the standard Dedekind–Cantor real line. In other places, we fuse some regions without bothering to check how many regions are so fused. We take this play with infinity in Chapter 2 to be "actual", since the (actual) existence of a *single* region—a part of the space—is established with each fusion.

The straightforward Aristotelian "fix" is to replace the unrestricted fusion principle with one allowing the fusion of any two (and thus any finite number of) regions. The resulting theory, however, is extremely weak, failing to prove the existence of meets, differences, bisections, and biextensions, and also failing to prove an Archimedean property. So those are added as additional axioms. It is perhaps interesting that adding a notably unAristotelian principle of unrestricted fusion allows one to prove several theorems that are, in fact, legitimate for the Aristotelian.

Building on some previous work by Linnebo, we then show how to take the "potential" nature of the usual operations seriously, by using a modal language (thus down-playing the fusion principle), and we show that modal and non-modal approaches are equivalent. We think that the modal approach better captures the matter of potentiality (but we make no attempt to explicate, let alone conform to, the often vexed account of modality proposed by Aristotle himself).

The chapter concludes with a predicative account of the continuum. Intuitively, that theory lies between the semi-Aristotelian account of Chapter 2 and the more Aristotelian account here. The predicativist accepts the existence of some infinite collections (or pluralities) of regions, those that can be defined in a predicatively acceptable way. The usual range of issues concerning predicative analysis re-emerge here.

Chapter 4 is an attempt to recover "points" and (something like) real numbers in the more Aristotelian framework presented in Chapter 3, along similar lines to the analogous development in the "semi-Aristotelian" Chapter 2. Getting analogues of the rational numbers is straightforward enough, but, we think, that exhausts Aristotelian resources.

As noted above, Aristotle and, just about everyone else until the nineteenth century, rejected the existence of the actual infinite. Ancient, medieval, and early modern mathematicians worked with specific procedures that can be iterated indefinitely often, thus introducing potential infinities. What they did not have, so far as we know, is a *theory of* potentially infinite sets or sequences. To put the matter a bit anachronistically, the ancient, medieval, and early modern mathematicians did not countenance bound variables that range over potentially infinite entities.

That had to wait for the advent of intuitionistic and constructive mathematics, in the twentieth century. The plan for this chapter is to follow their lead and develop a "superstructure" of "points" and "real numbers" on the gunky, fully Aristotelian framework, understood in terms of potentially infinite processes. It turns out that, unlike the foregoing semi-Aristotelian account, the punctiform superstructure here is very different from the underlying, gunky space—we do not find anything like mutual interpretability between the gunky and more punctiform theories. In most of the systems, we reproduce the constructive result that all functions on real numbers are continuous. It follows that the punctiform superstructure has a strong form of indecomposability: it is not possible to divide an "interval" cleanly into two parts. This is in contrast with the underlying, Aristotelian gunky framework which boasts a principle of bisection.

Again, in the semi-Aristotelian framework, the punctiform superstructure is isomorphic to the Dedekind–Cantor real numbers, and the Dedekind–Cantor theory is mutually interpretable and, indeed, definitionally equivalent to the gunky semi-Aristotelian one. We take this difference between the semi-Aristotelian and the more Aristotelian frameworks to be significant. This, too, highlights the significance of the actual infinite, especially when construed impredicatively.

The next order of business is to present higher-dimensional, regions-based theories (from the semi-Aristotelian perspective). The one-dimensional case is simplified somewhat by the feature that there are only two "directions" for the line, which we label "left" and "right". A two-dimensional space is much richer having, intuitively, infinitely many "directions".

Chapter 5 develops a Euclidean, two-dimensional, regions-based theory. As with the semi-Aristotelian account in Chapter 2, the goal here is to recover the now orthodox Dedekind–Cantor continuum on a point-free basis.

We first derive the Archimedean property for a class of readily postulated orientations of certain special regions, what we call "generalized quadrilaterals" (intended as parallelograms), by which we cover the entire space. Then we generalize this to arbitrary orientations, and then establish an isomorphism between the space and the usual point-based $\mathbb{R} \times \mathbb{R}$. As in the one-dimensional case, this is done on the basis of axioms which contain no explicit "extremal clause", and we have no axiom of induction other than ordinary numerical (mathematical) induction. After explicitly defining "point" and "line", as extensive abstractions, we derive the characteristic Parallel's Postulate (Playfair axiom) from our regions-based axioms, and point the way toward deriving key Euclidean metrical properties.

The extension from two to three and higher dimensions is fairly straightforward. We sketch how it can be accomplished.

Chapter 6 adapts the foregoing results to present two non-Euclidean theories, both in line with the (semi-)Aristotelian theme of rejecting points, as parts of regions (but working with actual infinity). The first theory is a two-dimensional *hyperbolic* space, that is, one that has a negative constant curvature. This space arises, in the standard point-based treatments, from adopting the Euclidean axioms for the plane except, of course, the Parallel's Postulate, which is replaced by an axiom stating that for any "line" l and "point" p, there is more than one "line" (usually infinitely many such "lines") l' through p that never meets l. Furthermore, the angle sums of polygons are less than their Euclidean counterparts, with the difference ("defect") shrinking as the polygons considered get smaller, approaching, but never reaching, the Euclidean values. Unlike the Euclidean case, the "shrinking" polygons cannot be similar.

Our second theory captures a space of constant *positive* curvature, a two-dimensional *spherical* geometry. The task here is to formulate axioms on regions which allow us to prove that (i) there are no infinitesimal regions and (ii) that there are *no* parallels to any given "line" through any "point" not on the given "line".

In both cases, of course, we have to define "point" and "line", since, in line with the theme of the book, regions do not contain points or lines, as parts. Our axioms, and definitions, imply the other Euclidean axioms, except, of course, the statement that any two points determine a unique line. In the spherical case, the exceptions are "antipodal points".

Chapter 7 turns to matters of metaphysics. Some analytic metaphysicians have occupied themselves with the nature (or the possible nature) of space and time (or space–time), and with the relationship between physical objects and the regions of space or space–time they occupy. Some of the issues concern the boundaries of objects and the notion of contact. Much, but not all, of the metaphysical literature just assumes, without comment, that the underlying space or space–time is punctiform, following the now orthodox Dedekind–Cantor picture. That is, many writers assume that space or space–time is composed of points—it is structured like $\mathbb{R}^3$ or $\mathbb{R}^4$. Physical objects thus occupy regions that are nothing but sets of points. Many of the authors get tangled up over what to say about point-sized regions, or even point-sized objects. Our first goal in this chapter is to give a somewhat biased overview of a portion of this literature, arguing that many of the issues are much easier to negotiate if we assume a regions-based space or space–time.

Some metaphysicians claim that physical objects can fully occupy a region—a set of points—only if the region is regular open (or regular closed). Those are the "regions" in the main topological models of our own theories. So those metaphysicians would have exactly the same picture if they began with a regions-based account of space, or space–time, like ours.

We then turn to some apparent limitations of our semi-Aristotelian accounts of space or space–time. For example, the natural analogue of Lebesgue measure is not countably additive (although it is finitely additive), and there seems to be no straightforward way to account for continuous variation in our frameworks—other than by just introducing "points" via "extensive abstraction", and working with those (as in the Dedekind–Cantor picture). Also discussed is the topic of differential geometry, of central importance in modern space–time physics. Rather than attempting a direct, regions-based theory mathematically equivalent to standard differential geometry (say, semi-Riemannian four-dimensional), we point out that our reduction of punctiform real analysis already is capable of incorporating such a theory, despite its sweeping complexity. Concluding this section, we discuss the extent to which our responses in the face of these challenges represent shortcomings of the regions-based, gunky approaches.

We then broach the question of adjudicating whether space or space–time really is punctiform. The tight connection between our regions-based, gunky theories and the more standard Dedekind–Cantor punctiform theories indicates that space or space–time can be described, completely and adequately either way. If one theory is adequate, then so is the other to exactly the same extent. This suggests that, given the contemporary notion of the actually infinite, a "dispute" between an advocate of points and one who rejects points is something like a Carnapian, external question. We support this with a treatment of verbal disputes.

Our study concludes with a brief sketch of different accounts of continuity: punctiform, intuitionistic, smooth, predicative, etc., and indicates the extent to which each captures several long-standing features that have been attributed to continuous entities. No one theory captures them all and, indeed, no one theory can capture them all. The leading intuitive properties of the continuous exhibit tensions with one another, and, we submit, pertain to a concept that is more up for sharpening in various incompatible ways, rather than one that would yield to more traditional, univocal philosophical analysis.

2

The Classical Continuum without Points

2.1 Atomless Mereological Continuum

The system developed in this chapter characterizes a one-dimensional continuum consisting of "regions" as parts, including "intervals". This space does not have any points as parts, although we will be able to define "point" in terms of intervals, a kind of Whiteheadian "extensive abstraction". So the notions of "open", "closed", and "half-open" are simply not available. Once we have proved that our continuum is Archimedean, we will demonstrate that it is isomorphic to the classical Dedekind–Cantor continuum, as a complete, separable, linearly ordered field.

Our formalism begins with classical first-order logic with identity supplemented with a standard axiom system for second-order logic (or logic of plural quantification, with an unrestricted comprehension axiom for plurals),[1] and with an adaptation of the standard (Tarskian) axioms of mereology together with (something implying) an "Atomless" axiom.

Axioms of Mereology:

1a. Axioms on $x \leq y$ **("**x **is part of** y**"): reflexive, antisymmetric, transitive.**

Certain of our axioms and theorems are conveniently stated in terms of a binary relation called "overlaps":

$$x \circ y \Leftrightarrow^{df} \exists z(z \leq x \,\&\, z \leq y).$$

We write $x|y$ for $\neg\exists z[z \leq x \,\&\, z \leq y]$, pronounced "$x$ is discrete from y (and vice versa)".[2]

1b. Axiom on $\leq$ **and** $\circ$**:** $x \leq y \leftrightarrow \forall z[z \circ x \rightarrow z \circ y]$.

[1] This looks very much like second-order logical comprehension for monadic predicates, except that it is conditional upon there being something satisfying the predicate. It may be written:

$$\exists v(\Psi(v)) \rightarrow \exists xx \forall y[y \prec xx \leftrightarrow \Psi(y)],$$

where '$y \prec xx$' is read "y is one of (or is among) the xx", and where Ψ is any formula of the language lacking free 'xx'.

[2] We prefer the term "discrete" to "disjoint" here in order to emphasize the fact that our regions are not sets of points (or anything else).

Theorem 1. *Axioms 1a and 1b imply the Extensionality Principle:*

$$x = y \leftrightarrow \forall z[z \circ x \leftrightarrow z \circ y].$$

Proof. From left to right is trivial. (Take y as x, then substitute y for the second x.) From right to left: Assume the right half of Extensionality and rewrite it as the conjunction of two conditionals: $\forall z[z \circ x \rightarrow z \circ y]$ & $\forall z[z \circ y \rightarrow z \circ x]$. By Axiom 1b, the first of these yields $x \leq y$, and the second yields $y \leq x$. By antisymmetry of $\leq$, the conjunction of these is equivalent to $x = y$. ■

So we adopt the so-called "classical" mereology.

2. **Fusion or whole comprehension:**

$$\exists u \Phi(u) \rightarrow [\exists x \forall y \{y \circ x \leftrightarrow \exists z(\Phi(z) \,\&\, z \circ y)\}],$$

where Φ is a predicate of the second-order language (or language of plurals) lacking free x.[3]

At this point, we could add an Atomless axiom: $\forall x \exists y(y < x)$, where $y < x \leftrightarrow^{df} y \leq x \,\&\, y \neq x$ (read "*y is a proper part of x*"). But this will follow from a stronger condition imposed below on the interval structure of our "pointless" or "gunky" line (Axiom 5).

We write $x+y$ for the mereological sum or fusion of x and y, so that $\forall z[z \circ x+y \leftrightarrow (z \circ x \vee z \circ y)]$, and we use $\sum_{n=0}^{\infty} x_n$ to designate fusions of infinitely many regions.

If $x \circ y$, then we write $x \wedge y$ for the *meet* of x and y. It is the fusion of all regions that are part of both x and y. So $\forall z[z \leq x \wedge y \leftrightarrow z \leq x \,\&\, z \leq y]$ (and if x and y have no common part, then $x \wedge y$ is undefined).

Similarly, if $\exists z(z \circ x \,\&\, \neg(z \circ y))$, then we let $x - y$ be the fusion of all regions z that are part of x but discrete from y (and if there is no such z, then $x - y$ is undefined). So $\forall z(z \leq x - y \leftrightarrow (z \leq x \,\&\, \neg(z \circ y)))$.[4]

By Axiom 2, fusions always exist, and meets and differences also exist wherever defined.

We label the pointless line we wish to characterize G, for "gunky". Below, we'll prove that (quite remarkably) our very elementary axioms suffice to characterize G precisely as a certain minimal closure; and then we'll prove that, with its interval structure, G is isomorphic to the classical real-numbers, $\mathbb{R}$. The point, of course, of having the Atomless condition is to ensure that, literally, G contains no points at all. Thus, the

[3] The formulation in the language of plurals takes this form:

$$\forall uu\{\exists w(w \prec uu) \rightarrow \exists x \forall y[y \circ x \leftrightarrow \exists z(z \prec uu \,\&\, z \circ y)]\},$$

where '*uu*' is a plural variable. If plural variables are assumed to have instances, then the antecedent and the main conditional can be omitted.

[4] The definitions of "meet" and "difference" given in [59], [60] are incorrect.

range of our first-order and plural variables can be thought of as all the parts of G, which we call "regions".

It is convenient to introduce a geometric primitive, $L(x, y)$, to mean "x is (entirely) to the left of y". The axioms for L specify that it is irreflexive, asymmetric, and transitive. And we define '$R(x, y)$', "x is (entirely) to the right of y", as $L(y, x)$.

Now we can introduce an important geometric relation, *betweenness*: $Betw(x, y, z)$ for "y is (entirely) between x and z":

$$Betw\,(x, y, z) \Leftrightarrow^{df} [L(x, y) \,\&\, R(z, y)] \vee [R(x, y) \,\&\, L(z, y)]$$

It follows that $Betw(x, y, z) \leftrightarrow Betw(z, y, x)$.

$L(x, y)$ obeys the following axioms:

3a. $L(x, y) \vee R(x, y) \rightarrow x|y$. (Of course, $x|y$ implies $x \neq y$.)
3b. $L(x, y) \leftrightarrow \forall z, u[\, z \leq x \,\&\, u \leq y \rightarrow L(z, u)]$.

The following can now be inferred:

$Betw(x, y, z) \rightarrow x|y \,\&\, y|z \,\&\, x|z$, and
$Betw(x, y, z) \,\&\, Betw(u, x, z) \rightarrow Betw(u, y, z)$,

where the transitivity of L is used for the latter.

Next we define an essential notion, that of a *connected* part of G. Intuitively, such a part has no gaps. The definition is straightforward:

$$Conn(x) \Leftrightarrow^{df} \forall y, z, u[z, u \leq x \,\&\, Betw(z, y, u) \rightarrow y \leq x]. \qquad \text{(Df } Conn\text{)}$$

In words, x is connected just in case anything lying between any two parts of x is also a part of x.

Furthermore, we can define what it means for a part p of G to be *bounded:*

$$Bounded(p) \Leftrightarrow^{df} \exists x, y\, Betw(x, p, y). \qquad \text{(Df } Bounded\text{)}$$

In words, a region wholly between two others is bounded.

We call bounded connected regions "*intervals*" and write '$Int(j)$', etc., when needed. However, note that, lacking points, we cannot describe intervals as either "open" or "closed", or "half-open".

Once we establish that G is bi-infinite, i.e. infinite in both directions, it will follow that, for connected regions, boundedness is a necessary condition for "*finite in extent*", as commonly understood, namely as finite relative to any interval. And once we have established that G is Archimedean, it will follow that boundedness is also sufficient for "finite in extent".

Using L, we can impose a condition of dichotomy for discrete intervals:

4. Dichotomy axiom: $\forall i, j[i, j$ are two discrete intervals $\rightarrow (L(i, j) \vee L(j, i))]$.

Now we can prove a linearity condition among intervals:

Theorem 2. *(Linearity): Let x, y, z be any three pairwise discrete intervals; then exactly one of x, y, z is between the other two.*

Proof. Applying Dichotomy to the hypothesis, assume that (say) $L(x, y)$. If also $L(y, z)$, then $R(z, y)$, so that $Betw(x, y, z)$, and this is unique. If instead $L(z, y)$, then either $L(z, x)$, in which case we have $Betw(z, x, y)$, uniquely; or $L(x, z)$, in which case we have $Betw(x, z, y)$, also uniquely. The argument assuming at first that $R(x, y)$ is similar. ■

To guarantee that arbitrarily small intervals exist everywhere along G, we adopt the following axiom:

5. $\forall x \exists j[Int(j) \,\&\, j < x]$.[5]

An important relation of two intervals is "adjacency", which is defined as follows:

$$Adj(j, k) \Leftrightarrow^{df} j|k \,\&\, \nexists m[Betw(j, m, k)]. \qquad \text{(Df } Adjacent\text{)}$$

As noted in Chapter 1, this is Aristotle's notion of contiguity.

Now suppose that $j = \Sigma_{i=1}^{\infty} j_i$, where $Int(j_i)$ and $R(j_{i+1}, j_i)$ and $Adj(j_{i+1}, j_i)$. Our axioms on L imply that $R(k, j)$ just in case $\forall i R(k, j_i)$, and analogously for $L(k, j)$.

The following equivalence relations on intervals will also prove useful: "*j and k are left-end-equivalent*" just in case

$$\exists p[p \leq j \,\&\, p \leq k \,\&\, \nexists q(\{q \leq j \vee q \leq k\} \,\&\, L(q, p))].$$

Left-end-equivalence of *j* and *k*:

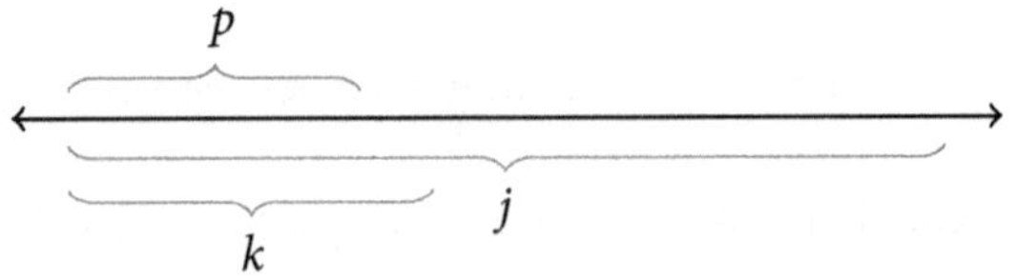

"*Right-end-equivalent*" is defined analogously. Left-(Right-)end-equivalence means, intuitively, that the intervals "share their left (right) ends, or endpoints, in common", as in Aristotle's conception of two things being continuous with each other. However, our system does not recognize "ends" or "endpoints". So we need a definition like the above.

One further geometric primitive is very useful both in ensuring that G is infinite in extent and in recovering, in effect, the rational numbers as a countable, dense subset of the (arithmetic) continuum, viz. *congruence*, as a binary relation among intervals. Intuitively, $Cong(i, j)$ is intended to mean "the lengths of intervals i and j are equal". Thus, we adopt the usual first-order axioms specifying that *Cong* is an equivalence relation. We will sometimes write this as $|i| = |j|$, but with the understanding that

[5] This of course implies the "Atomless axiom", introduced above. If one were to add a "bottom" element (an empty region), the result would be an atomless, complete Boolean algebra.

we have not yet given any meaning to '$|i|$' standing alone, but only in certain whole contexts. Similarly, for intervals i, j, we can define, contextually, $|i| < |j|$ as meaning: $\exists j'[j'$ an interval & $j' < j$ & $\text{Cong}(i, j')]$; and we may write $|i| > |j|$ as equivalent to $|j| < |i|$.

We come now to a key axiom, crucial to our characterization of G:

6. Translation axiom: Given any two intervals, i and j, each is congruent both to a unique left-end-equivalent and to a unique right-end-equivalent of the other.

In effect, this guarantees that a given length can be "transported" (more accurately, instantiated) anywhere along G, and that these instances are unique as congruent and either left- or right-end-equivalent to the given length.[6] In particular, we can prove

Lemma 1. *Given any two intervals i and j such that $\neg Cong(i, j)$, either there exists an interval $i' < j$ with $Cong(i, i')$; or there exists i' with $j < i'$ with $Cong(i, i')$.*

Proof. By $\neg Cong(i, j)$, $i \neq j$. Assume that $\neg(i' < j)$ for any i' such that $Cong(i, i')$. By the Translation axiom, there exists i' such that $Cong(i, i')$ and i' is left-end-equivalent to j.

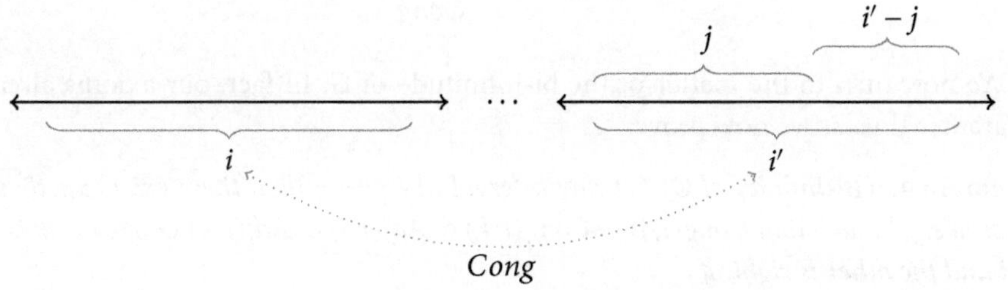

We want to show that $j \leq i'$, as that will establish that $j < i'$, as desired. Assume the contrary, i.e. that $j \nleq i'$. Now, if $i' - j$ doesn't exist, then, by definition, $i' \leq j$. But, since $Cong(i, i')$, we have $\neg Cong(i', j)$, whence $i' \neq j$, and then we would have $i' < j$, contrary to hypothesis. So assume some $n \leq i' - j$. By the hypothesis for *reductio*, there is also $k \leq j$ & $k \nleq i'$, and indeed $\neg(k \circ i')$. Without loss of generality, we may assume that k is an interval. (See Axiom 5.) Since i' is left-end-equivalent to j, it follows that k is *not* left-end-equivalent to j. But there is an $m \leq j - k$ which is left-end-equivalent to both j and i' so satisfies $m \circ i'$. Let m' be a common part of m and i'. Clearly $L(m', k)$. But $\neg(L(n, k))$, since if it were, it would overlap j, contrary to assumption. (n can't be left of j, since it's part of i', and $i'$and j are left-end-equivalent.) Therefore, by the Dichotomy axiom on L, we have $L(k, n)$, whence $Betw\,(m', k, n)$, with both $m', n \leq i'$ but $k \nleq i'$, contradicting that i' is an interval. ■

[6] Euclid sometimes invokes a two-dimensional analogue of "translation". For example, Proposition 4 of Book 1 is proved by moving one triangle on top of another. Euclid does not give a postulate that justifies this.

Theorem 3. *(Trichotomy) For any two intervals, i, j, either $|i| = |j|$ or $|i| < |j|$ or $|i| > |j|$.*

Proof. Immediate from Lemma 1 and the definitions of the disjuncts. ∎

One further axiom on congruence is useful and intuitively intended, viz. that congruence respects nominalistic summation of adjacent intervals:

7. Additivity: Given intervals i, j, i', j' such that $Adj(i,j)$, $Adj(i',j')$, $Cong(i,i')$, $Cong(j,j')$, then $Cong(k,k')$, where $k = i + j$ and $k' = i' + j'$.

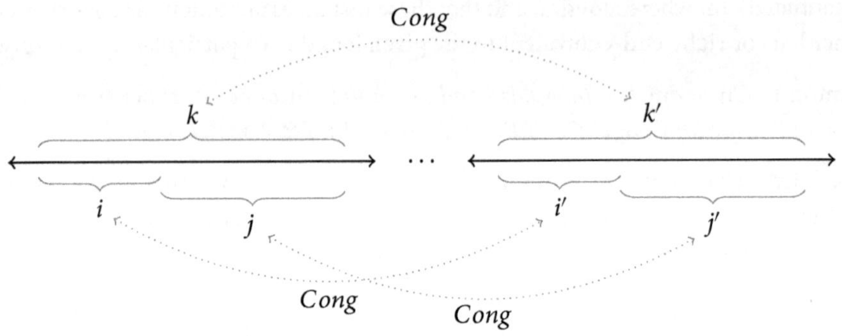

We now turn to the matter of the bi-infinitude of G. In fact, our axioms already guarantee this, as we now prove.

Theorem 4. *(Bi-Infinity of G) Let any interval i be given; then there exist exactly two intervals, j, k, such that Cong(i, j) & Cong(i, k) & Adj(i, j) & Adj(i, k) & one of j, k is left of i and the other is right of i.*

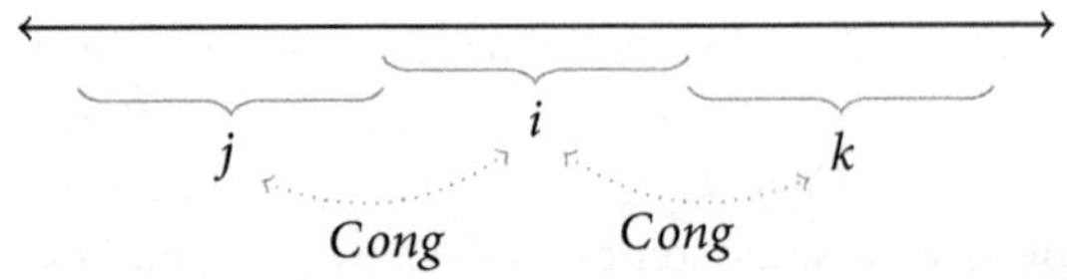

Proof. Given an interval i, by definition it is bounded, so there exist regions that are left of i and regions that are right of i. Assume a region m to (say) the right of i. (The case for m to the left is handled exactly analogously.) If $Adj(m,i)$, then, by the Translation axiom, there is a unique interval j such that $Cong(j,i)$ and j is left-end-equivalent with m. If not-$Adj(m,i)$, then let f be the fusion of all intervals p such that $Betw(i,p,m)$. f is an interval. Then, by Translation, there is a unique interval j with $Cong(i,j)$ and j left-end-equivalent to f. Combining this with the analogous argument for the case with m to the left of i completes the proof. ∎

Since biextension obviously iterates, this already ensures that G is "bi-infinite" in the sense of containing, as a part, the fusion of the minimal closure of any interval i under the operation of "*biextension*" defined in the theorem. (This closure is proved to exist in Lemma 3.) But we can do better and also ensure that G is *exhausted* by iterating the process of flanking a given interval by two congruent ones as in Bi-infinity. This is just the Archimedean property, derived below. Toward this end, call an interval l an (*immediate*) *biextension* of interval i—$BiExt(l, i)$, or $biext(i) = l$—just in case $l = j + i + k$, where j, i, k behave as in the Bi-infinity theorem.

Lemma 2. *Let i and j be intervals such that $i < j$; then $\neg Cong(i, j)$.*

Proof. For a contradiction, assume $Cong(i, j)$. There are three possible cases: (1) i is left-end-equivalent to j; (2) i is right-end-equivalent to j; (3) i is neither. Cases (1) and (2) are argued in exactly the same way. For definiteness, assume case (1). By Bi-infinity, there exists i' extending i to the left with $Cong(i', i)$ and $Adj(i', i)$, hence $Adj(i', j)$.

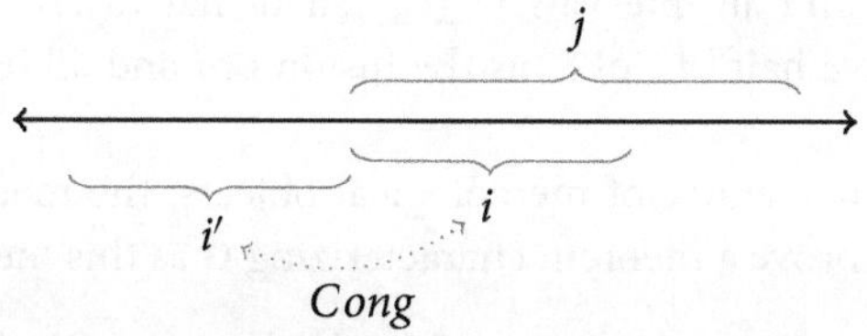

But then, by the hypothesis for reductio and transitivity of *Cong*, it follows that both i and j qualify as the unique right extension of i', as required by Bi-infinity, and since, by hypothesis of the Lemma, $i \neq j$, this is a contradiction. Case (2) is argued exactly analogously, considering i' as extending i to the right.

In case (3), let k_L be the fusion of all parts x of j such that $L(x, i)$ and let k_R be the fusion of all parts x of j such that $R(x, i)$. k_L and k_R are intervals. (Easy exercise.) Clearly, $k_L + i + k_R = j$ and this sum is discrete (all three pairs discrete). Now, let j' be (say) the right extension of j, i.e. $Cong(j', j)$ and $Adj(j', j)$. By Translation, let i' satisfy $Cong(i', i)$ with i' left-end-equivalent to j'. By Translation again, let k'_L satisfy $Cong(k_L, k'_L)$ and $Adj(i', k'_L)$ with $L(i', k'_L)$; and let k'_R satisfy $Cong(k_R, k'_R)$ and $Adj(k'_L, k'_R)$ with $L(k'_L, k'_R)$.

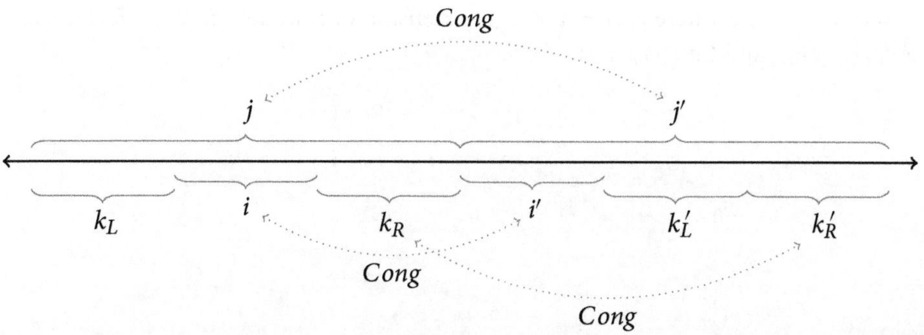

Then by Additivity, $Cong(j, i' + k'_L + k'_R)$, so, by the uniqueness of (right) extension of j as required by Bi-infinity, we have $j' = i' + k'_L + k'_R$, whence $i' < j'$, whence $i' \neq j'$, but then both j' and i' qualify as the unique right extension of j, a contradiction. ■

Now we can characterize G. Toward that, let X be any class of intervals such that an arbitrary but fixed interval $i \leq G$ is one of the X and such that if $k = biext(j)$ for j any of the intervals of X, then k is also in X. Call such X a "closure of i under *biext*".

Lemma 3. By Axiom 2, there is an individual which is the common part of the fusions of each class X which is a closure of i under *biext*, which we call their *meet* or *the minimal closure* i^* *of i under biext*. (Since i is stipulated to belong to any such X, the meet is non-null, as required in mereology.)

Proof. Immediate from Axiom 2. ■

Given a fixed "unit" interval, i, we define the "right-half" or "positive half" i^+ of i^* as the fusion of i and all intervals $j \leq i^*$ such that $R(j, i)$. Then we define the "left-half" or "negative half" i^- of i^* as the fusion of i and all intervals $j \leq i^*$ such that $L(j, i)$.

By the criterion for identity of mereological objects, the meet i^* of Lemma 3 is unique. We now can prove a theorem characterizing G as this meet:

Theorem 5. *(Characterization of G): Let G be the fusion of the objects in the range of the quantifiers of our axioms; then* $G = i^*$, *the fusion of the minimal closure of i under biext.*[7]

Proof. Suppose, for a contradiction, that $G \neq i^*$. Since, by stipulation, $i \leq G$ and G is closed under *biext*, we have that $i^* < G$. Then some part p, indeed (by Axiom 5) an interval $k \leq G$ satisfies $\forall j[Int(j) \,\&\, j \leq i^* \rightarrow L(k, j)] \vee \forall j[Int(j) \,\&\, j \leq i^* \rightarrow R(k, j)]$; therefore, by definition, $L(k, i^*) \vee R(k, i^*)$. Let's suppose it's $R(k, i^*)$. (The other case is argued exactly analogously.) Recall that i^+ is the positive or right half of i^*, as defined above. Clearly i^+ is connected; and by our betweenness criterion, it is also "bounded", so an interval. Therefore, by the Translation axiom, there is a unique interval $m \leq i^+$ with the properties (1) m is right-end equivalent to i^+, and (2) $Cong(m, i)$. But this leads to contradiction, as follows: By definition of i^+ as the positive "half" of i^*, the minimal closure of i under biextension, we have that i^+ is the fusion of a sequence $\langle i_n \rangle$ where $i_0 = i$, and where $i_{n+1} =$ the right extension of i_n under *biext*, i.e. $R(i_{n+1}, i_n)$, $Adj(i_{n+1}, i_n)$, and $Cong(i_{n+1}, i)$.

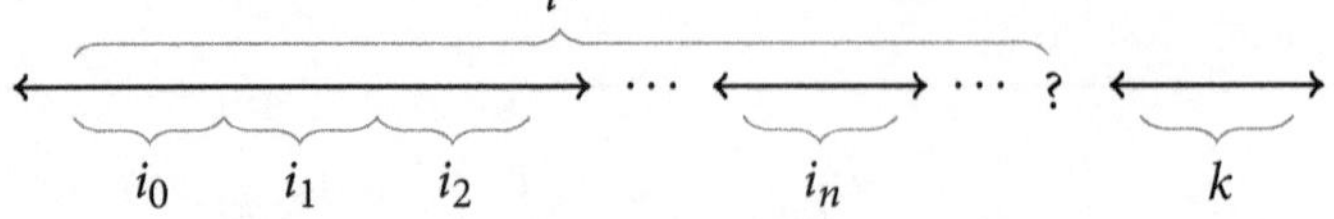

[7] A closely analogous argument rules out infinitesimal intervals (relative to a given unit interval i). We suggest that these results, expressing that G is Archimedean, are quite surprising, since no axiom explicitly contains an "extremal clause" to the effect that the intervals of G are *only* those that are part of the fusion of those obtained by repeated applications of *biext* starting with a given interval. Nor do we have an induction axiom for properties of intervals, although, of course, in light of Theorem 3, such an induction principle could be derived from properties of minimal closures, à la Frege.

Now, since $m \leq i^+$, $m \circ i_k$, for some k. Therefore, there is a least such k, call it k_0, such that $m \circ i_{k_0}$. But then, for any $k > k_0$, $i_k < m$. But both i_k and m are congruent to i, so this contradicts Lemma 2. This shows that the assumption of such m, hence of a $k \leq G - i^*$, must be wrong. Thus, taking account of the exactly parallel argument for the left (negative) half of i^*, it follows that $G = i^*$. ■

Thus, our axioms on the interval structure of G already imply the Archimedean property directly, without a detour through the point-based continuum, $\mathbb{R}$. Indeed, the same can be said of Dedekind completeness, as is shown in Appendix 2.1.

Finally, we need a guarantee that any interval has a unique bisection. That can now be proved as a theorem:

Theorem 6. *(Existence and uniqueness of bisections): Given any interval i, there exist intervals j, k such that* $j < i$ *&* $k < i$ *&* $j|k$ *&* $j + k = i$ *& Cong*(j, k)*; and j, k are unique with these properties.*

Proof. Let i be any interval. For any interval j, let j^{+r} be the fusion of j and the right biextension of j. So we need to find an interval j that is left-end-equivalent with i such that $i = j^{+r}$. Let k be any interval such that $k < i$. Without loss of generality, assume that k is left-end-equivalent with i. We have that $i - k$ exists, and is an interval.

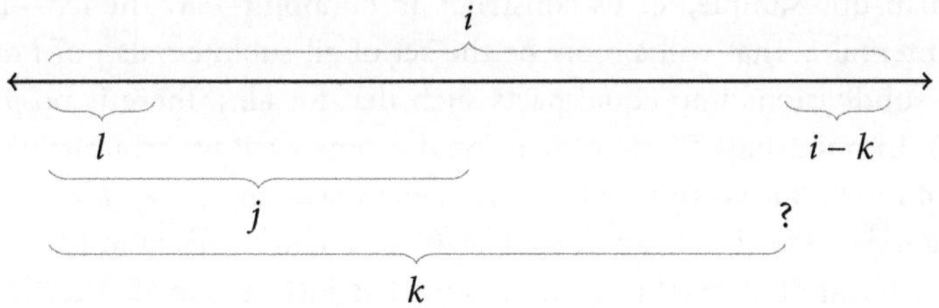

If k is congruent with $i - k$, we are done. So suppose not. Either $|k| < |i - k|$ or $|i - k| < |k|$. Let l be an interval that is congruent to the smaller of those two and is left-end-equivalent with i. So $l^{+r} \leq i$. (In fact $l^{+r} < i$.) Now let j be the fusion of all intervals m such that m is left-end-equivalent with i and $m^{+r} \leq i$. Clearly $j^{+r} \leq i$. If $j^{+r} = i$, we are done. So suppose that $j^{+r} < i$. Let $n = i - j^{+r}$. Then n is an interval. Let p be an interval such that $p < n$ and, without loss of generality, suppose that p is left-end-equivalent to n. Let q be $n - p$. Without loss of generality, assume that either $|p| = |q|$ or $|p| < |q|$. So $p^{+r} \leq n$. Let j' be the fusion of j and an interval congruent to p immediately on its right. (First let p_1 be congruent with p and right-end-equivalent with j. Then let p_2 be the right biextension of p_1.Then $j' = j + p_2$.)

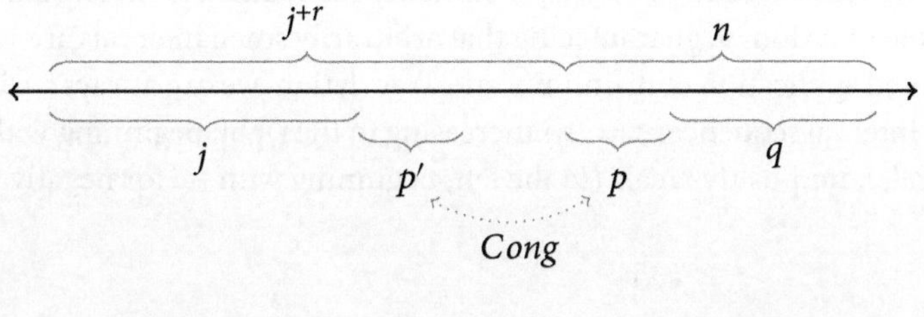

An application of Additivity shows that $j'^{+r} = j^{+r} + p^{+r}$, and we have $j^{+r} + p^{+r} \leq i$. This contradicts the definition of j as the fusion of all intervals m such that m is left-end-equivalent with i and $m^{+r} \leq i$.

For uniqueness, given interval i, suppose both $i = j_L + j_R$ with $Cong(j_L, j_R)$ and $Adj(j_L, j_R)$, and also $i = k_L + k_R$ with $Cong(k_L, k_R)$ and $Adj(k_L, k_R)$, with neither $j_L = k_L$ nor $j_R = k_R$. Suppose without loss of generality that $k_L < j_L$(and so $j_R < k_R$). Let $m = j_L - k_L$. Then $j_R + m = k_R$, whence $Cong(j_R + m, k_L)$. Now let m' be congruent to m and adjacent to m to its right. Then $Cong(j_R - m', k_L)$. But we have $j_R - m' < j_R < j_R + m = k_R$ (where j_R is discrete from m). Since $Cong(j_R - m', k_R)$, this contradicts our lemma that if $i < j$, then $\neg Cong(i, j)$. ∎

2.2 Recovering $\mathbb{R}$

By repeated application of bisections, we can, in effect, approximate any locus along G to within any desired accuracy with sufficiently many nested intervals, whose least after k subdivisions is of norm 2^{-k} assuming the initial "unit interval" i is of norm 1. (Here we are speaking in our metalanguage, not yet having reconstructed the norm function in our object language.) One natural strategy that now suggests itself is to *define* an exact locus or "point" as a "Cauchy sequence" of such decreasing intervals.

As a warm-up example, let us construct an endpoint—say the left—for a given arbitrary interval i. That will simply be the set of all subintervals j of i obtained by successive subdivisions into equal parts such that for all j, there is no $p < i$ such that $L(p, j)$. In point-based, one-dimensional geometry, if we arbitrarily set the left-endpoint of $i = 0$, this corresponds to the Cauchy sequence: $\langle \frac{1}{2^k} \rangle$, $k = 1, 2, \ldots n \ldots$, converging to 0. Indeed, we can introduce '0' (as a *real number*) in exactly this way: let i be an arbitrary but fixed interval, oriented as just described. Then $0 =^{df} \cap[S :$ S contains i and contains the left half of any subinterval j of i such that $j \in S]$. Similarly, we could define 1, replacing 'left' (L) with 'right' (R). (Below, however, we give a definition in terms of Cauchy sequences of intervals increasing to the right, in conformity to the rest of the positive half of G.) Notice that these "objects", whether thought of as numbers or as points, are not claimed to be parts of G. On the contrary, they are part of a meta-theoretic *superstructure* that we construct over the mereological-interval structure of G.

In general, we define a sequence $\langle j_i \rangle$ of intervals increasing to the right (or left, for negative reals) to be *Cauchy* just in case, for any interval, ε, there exists N such that for any $m > k > N$, $j_m - j_k$ is an interval, $R(j_m - j_k, j_k)$, and $|j_m - j_k| < |\varepsilon|$. (This last expression was defined contextually above. By assumption that $\langle j_i \rangle$ is increasing to the right, $j_m - j_k$ exists and $R(j_m - j_k, j_k)$. Similarly for sequences increasing to the left. Note the role of Axiom 5, guaranteeing that arbitrarily small intervals are values of ε.)

By repeated application of Bi-infinity and Translation, we can always avail ourselves of Cauchy interval sequences $s = \langle s_j \rangle$ increasing to the right, beginning with our fixed unit interval, i, for positive reals (to the left, beginning with $-i$ for negative reals), i.e.

such that $R(s_{k+1} - s_k, s_k)$ (and $L(s_{k+1} - s_k, s_k)$ for negative reals); the fusion of all the intervals s_j forms an interval.[8]

This proves convenient in giving a second representation of real numbers as intervals in G itself, which in turn—as we shall see below—augments the reach of reconstructions that don't rely on set theory. Thus, we will have available two relative interpretations of the classical continuum, $\mathbb{R}$, based on G: (1) equivalence classes of Cauchy sequences of intervals of G, or canonical ones from each class; and (2) fusions of canonical Cauchy interval sequences, as just indicated.

We'll return to (2) below. First, let us pursue (1) in some more detail. We want to construct the reals over the fixed interval i already associated with [0,1]. The first step is to identify the binary rationals as the appropriate subintervals left-endpoint-equivalent to i, obtained by iterated subdivisions licensed by the Bisection axiom. Thus, each rational of the form $\frac{n}{2^k}$, where $n = 1, 2, \ldots, 2^k - 1$, corresponds 1-1 with the left-endpoint-equivalent subintervals of i determined by the kth stage of bisections. (The reader will have noted that the full binary tree of Baire space is in effect generated by these subdivisions.) The next step is to identify arbitrary reals in $(0, 1]$ with increasing Cauchy sequences of these subintervals.[9]

The natural ordering of the binary subintervals of i implicit above is this: $j \prec k \leftrightarrow^{df} k - j$ exists (is non-null) & $R(k - j, j)$.

The next step is to extend this ordering to the increasing Cauchy sequences of intervals. We set $\langle r_i \rangle \prec \langle s_i \rangle$ just in case $\exists\varepsilon\exists N\forall k > N$ [ε an interval & $|s_k - r_k| > |\varepsilon|$ & $R(s_k - r_k, r_k)$] (Recall that the norm-notation was defined contextually using *Cong*, above.)

The pieces are now in place to prove a first recovery theorem:

Theorem 7. *The ordered structure of binary intervals within i together with the Cauchy sequences of them is order-isomorphic to the classical real numbers of* $(0, 1]$ *(in their natural ordering,* $<$*).*

Proof. There is a 1-1 invertible map φ from the binary intervals of i to the binary rationals of $(0, 1]$. Define φ as follows: Set φ of the left interval, call it $\frac{i}{2}$, of i resulting from the 1st subdivision $= \frac{1}{2}$; after the k'th subdivision, set φ of the left-most $= \frac{1}{2^k}$, of the next left-most $= \frac{2}{2^k}, \ldots$, of the next to right-most $= \frac{2^k - 1}{2^k}$. (φ of the right-most is of course always $= 1$.)[10] Clearly φ is order-preserving. To extend this to the increasing Cauchy interval sequences, map each such, of the form

[8] Since we don't recognize a null interval, 0 is conventionally defined either as above or as a rightward proceeding (nested) sequence starting with an interval, call it $-\frac{i}{2}$, congruent to the left half of i, choosing at each stage, k, the right–half subinterval of $-\frac{i}{2^{k-1}}$.

[9] Of course, sequences may omit both left and right subintervals at a given stage, corresponding to a '0' in the binary numeral representation of the real in question.

[10] Since e.g. $\frac{2}{2^k} = \frac{1}{2^{k-1}}$ etc., it appears that φ is many-one; but really it isn't, as the result of proceeding stepwise to the right simply adds a congruent interval adjacent to the preceding, so that the result is an interval, and in the case of an even number of steps, it is always equal to an interval obtained at an earlier

$\langle s_1, s_2, \ldots, s_n \ldots \rangle$ to the increasing rational Cauchy sequence determined by φ, viz. $\langle \varphi(s_1), \varphi(s_2), \ldots, \varphi(s_n), \ldots \rangle$. Call this extension of φ φ'. That φ' is 1-1 and onto the increasing binary rational Cauchy sequences is immediate from the properties of φ. That φ' is order-preserving is also clear: if $\langle r_k \rangle \prec \langle s_k \rangle$, then beyond some N (given in the definition of this ordering, above), $\exists n$ such that the corresponding rational differences, $\varphi(s_m) - \varphi(r_m) > 2^{-n}$, for any $m > N$, which defines order on these rational sequences. ■

Now we can extend this to the whole positive half-line, $(0, \infty)$ by applying the same procedure to right-extensions of i by any number of intervals each congruent with i. E.g. we map the interval $i + j$, where $Cong(i, j)$ & $Adj(i, j)$ & $R(j, i)$ to $(0, 2]$ (appealing to the Bi-infinity theorem), iterating this procedure to cover all intervals of the form $(0, n]$. Thus, we have:

Corollary 1. *The theorem statement holds for all intervals of the form* $(0, n]$, *hence for the whole positive part* G^+ *of* G *as order-isomorphic to the positive reals,* $(0, \infty)$.

Proof. The only thing to check, in addition to what has already been established, is that the map just introduced—call it φ''—from G^+ to $(0, \infty)$ is indeed defined on all of G^+ in the sense that no part $p \leq G^+$ is discrete from all the intervals on which φ'' is defined. Suppose, to the contrary, there is some such part, p. Then p must be discrete from each binary rational interval left-end-equivalent to i, and in particular discrete from any of the intervals obtained from i by any finite number of applications of the *biext* operation introduced above. But then p would be discrete from the fusion of all such intervals, which fusion = exactly the meet (= minimal closure of i under *biext*) that we proved = G in Theorem 5, above. This contradicts that $p \leq G^+$. ■

Corollary 2. *The theorem statement holds for all intervals of the form* $[-n, 0]$, *hence for the whole negative part* G^- *of* G *as order-isomorphic to the non-positive reals,* $(-\infty, 0]$.

Proof. Applying Translation and Bi-infinity, the constructions for the positive part of G can be shifted accordingly. (For the sake of the field operations, introduced below, it is simplest to reflect the positive intervals "about 0", i.e. proceeding leftward starting with $-i$ or subdivisions thereof.) The proof that the "mirror image" of the map φ'' is defined on all of G^- is exactly analogous to the proof of the first Corollary that φ'' is defined on all of G^+. ■

To extend the isomorphisms of the Corollaries to cover the whole of G, simply stipulate that, for every interval j of the negative part of G and every interval k of the positive part, $j \prec k$. Thus we have established:

Theorem 8. *G (i.e. $(G, \prec)$) is order isomorphic to $\mathbb{R}$ (ordered by $<$).*

Call the isomorphism of the latter theorem Φ.

subdivision. E.g. in one rightward step at the kth subdivision, we get left-most-$\frac{i}{2^k}$ + next-left-most-$\frac{i}{2^k}$ = left-most-$\frac{i}{2^{k-1}}$, etc. (where the '+' here is mereological summation).

With respect to the field-algebraic structure of $\mathbb{R}$, we can proceed in either of two ways. (1) We could regard it as additional structure of $\mathbb{R}$, built up in the usual way from the Cauchy sequences defining the reals, not bothering about any additional structure of G; or (2) we can introduce operations of "addition" and "multiplication" of intervals of G and prove that the order-isomorphism Φ is also an algebraic isomorphism. (2) is more interesting so let us pursue it.

To define an interval sum operation, call it $i \oplus j$, we can first define this for binary intervals, which we can already express as $\Phi^{-1}(q)$, for q a binary rational of $\mathbb{R}$, where such an interval is either left-end equivalent to i, if $q > 0$, or right-end-equivalent to $-i$, if $q < 0$, where $-i$ is the interval satisfying $Cong(i, -i)$ & $Adj(i, -i)$ & $R(i, -i)$. Then $\oplus$ will simply be vector-addition along G. $\oplus$ is then extended to all of $\mathbb{R}$ by applying it "pointwise" to inverse images under Φ of the (binary) rational Cauchy sequences defining the reals in question. To obtain "multiplication" of intervals, $i \otimes j$, either of two methods may be used. Remaining entirely within G, we first define this product for inverse images of binary rationals as iterated interval-addition, $\oplus$, treating binary fractions of intervals in accordance with the distributive law of multiplication over addition.[11] Then $\otimes$ is extended to all of $\mathbb{R}$ by applying it "pointwise" to inverse images under Φ of the rational Cauchy sequences defining the reals in question. Alternatively, we can adapt the established Euclidean geometric method of introducing the product of two lengths by working in the Euclidean plane. Here, of course, we work in $G \times G$, diagrammatically representing one copy of G as our abscissa, the other as ordinate, such that the left end of our unit, i, is the origin, where the two axes cross.[12] Then the product $j \otimes k$ is obtained by taking j as left-end-equivalent to i along the abscissa, k as extending from the origin along the ordinate (either "up" if we are operating on $+k$, or "down" if we are operating on $-k$). Next we construct the "hypotenuse" segment σ connecting the right-end of i with the "top"-end of k (if we're considering $j \otimes +k$, "bottom"-end of k if it's $j \otimes -k$). The value of the product is then represented as the interval m along the ordinate from the origin to where the segment, call it σ', meets the ordinate, where σ' extends from the right-end of j and lies parallel to σ. (σ' forms a second "hypotenuse", so we have two similar triangles. Then the definition of product derives directly from the fact that $\frac{|m|}{|k|} = \frac{|j|}{1}$.)[13] Now the definition of the operation $\otimes$ is extended to all of $\mathbb{R}$ via rational Cauchy sequences, as in the first method.

[11] e.g. $\Phi^{-1}(2.5) \otimes \Phi^{-1}(3.25)$ is computed by vectorially adding twice the (unit) interval i and half-i, call this $2.5i$, and then vectorially tripling this and adding a quarter of $2.5i$ to obtain the answer.

[12] Justifying this method on the present point-free basis would require adopting some further primitives and axioms to extend our methods to the Euclidean plane. We would need further equivalence relations of "end-equivalence" to replace reference to "the point where two non-collinear intervals meet"; and we would need a relation of angle-congruence for purposes of constructing paralells. All this will be carried out in Chapters 5–6 on two-dimensional continua.

[13] Note that we're now in a position to define the norm, $|j|$, for any interval j, based on the isomorphism Φ.

Recalling that $\Phi^{-1}(0)$ was introduced via a rational-interval Cauchy sequence, one checks that the algebraic laws of the field operations are preserved under Φ. Thus, we have:

Theorem 9. *G $(G, \prec, \oplus, \otimes)$ is ordered-field isomorphic to R $(R, <, +, \cdot)$.*

Although these isomorphism theorems rely on the characterization theorem (5) of G, which already expresses that G is Archimedean, it is worth pointing out that Theorem 9 implies that G satisfies a statement of the Archimedean property closer to the one commonly given for $\mathbb{R}$. Here is one way to formulate this. As will be suggested below, in Appendix 2.3, assume that our mereological universe is augmented with a natural-numbers-structure, N, constituted entirely of atoms, which we designate $0_N, 1_N, 2_N, \ldots$.[14] Then we can speak of functions from N $(', +_N, \bullet_N)$ to intervals of G by quantifying plurally over (unordered) pairs, $(n, j) =^{df} n + j$ (where this is nominalistic summing). Now inductively define a map $\varphi : N \to$ Intervals of G via

(*i*) $\varphi(0_N) = i$
(*ii*) $\varphi(n_N +_N 1_N) = biext(\varphi(n_N))$.

Now define 'i divides k (mod 3)' to mean: $k = \varphi(m_N)$, for some m_N. Then we say G *is Archimedean* just in case

$$\forall j \exists h [j \text{ an interval} \rightarrow h \text{ an interval} \ \& \ i \text{ divides } h \ (\text{mod } 3) \ \& \ j \leq h].$$

Thus we have

Corollary 3. *G is Archimedean.*

Proof. This follows from the Characterization Theorem on G as the minimal closure of i under *biext*. For a contradiction, suppose that some interval $j \leq G$ is not covered by any interval h obtained by iterating *biext*, starting with i, any finite number of times. Then, by construction of G, for some $j' \leq j$, $j'|G$, so $j \nleq G$, contradiction. (Cf. the proofs of Corollaries 1 and 2, via the definedness of the maps described there.) ■

Finally, we would like to be assured that the above development of $\mathbb{R}$ over G is independent of the starting interval, i, i.e. that starting with any other, j, leads to essentially the same recovery theorems, and even that the minimal closure of j under *biextension* is indeed G itself. That can be arrived at as follows.

Let j be any interval of G other than i. Now minimally close j under *biextension*, calling the result j^*. Then we proceed in two steps: (1) We can carry out the whole of the above construction to produce an ordered-field isomorphism Φ' from the binary intervals and Cauchy sequences thereof between G' and $\mathbb{R}$, where G' and Φ' are introduced just as G and Φ were but substituting reference to j for that to i throughout. (Think of a transformation of the interval structure of G based on i combining a

[14] This can itself be carried out along the lines of [56] or [57], ignoring modal operators for present purposes.

suitable translation and either a shrinking, in case $|j| < |i|$, or a stretching, in case $|i| < |j|$, or neither, if $Cong(i, j)$.) Step (2) consists of demonstrating that, in light of (1), the point-free continuum $G' = j^*$ based on j is indeed $= G$. That makes essential use of the Archimedean property of G', afforded by carrying out the proof of the last Corollary with G' in place of G, *mutatis mutandis.*

Theorem 10. *The point-free continuum* $G' = j^*$ *based on interval* j ($\leq G$ *and* $\leq G'$) $= G$.

Proof. By the definition of G as the minimal closure of i under *biextension*, $G' \leq G$. For since, by hypothesis, $j \leq G$, we have that for some finite n, $j \leq biext^n(i)$, i.e. j is part of the result of the nth iterate of *biext* applied to i. Then it is straightforward to show that any k covered by finitely many applications of *biext* starting with j is also covered by finitely many applications of *biext* starting with i. We need to show the converse, that $G \leq G'$. First we claim that $i \leq G'$. This follows from the Archimedean condition displayed above, applied to G', interchanging the roles of 'i' and 'j' to produce an h ($\leq G'$) such that $i \leq h$. Next we argue by induction that any interval k obtained from i by repeated application of the *biextension* operation will be accessible from j in the same sense as i is, i.e. by appealing again to the Archimedean condition above for G', substituting 'k' for 'j' and 'j' for 'i'. Then, from the definition of G, since every interval m of G is covered by the result of some finite number of iterations of *biextension* based on i, it follows from the induction that $m \leq G'$, as well, whence $G \leq G'$, and therefore $G' = G$. ■

2.3 Topological Models

Thus we recover the now standard Dedekind–Cantor continuum in our gunky space. We now go in the opposite direction, and present two topological models for our axiomatization in the standard real numbers. These further illustrate some of the Aristotelian notions of contiguity and continuity, and they show that our account is equivalent to the Dedekind–Cantor one in a strong sense. The exercise will also serve to remove any lingering doubts concerning the consistency of our axiomatization, if there are any.

For both models, the background meta-theory is the ordinary, Dedekind–Cantor account of the real numbers, with their usual topological properties. An open set S of real numbers is said to be *regular* if S is identical to the interior of its closure.[15]

[15] Richard Cartwright [31] argues that in 3-space, all and only regular open sets are "receptacles", regions of space that physical objects can occupy. Cartwright's theory is at least partly Aristotelian. This theme is developed further in Chapter 7, on metaphysics. Stefano Borgo [23] points out that most point-free geometries have models in the regular sets.

Define a real number r to be an *interior boundary* of a set S, if $r \notin S$, but there are numbers s, t such that $s < r < t$ and the open set $(s, t) - \{r\}$ is a subset of S. So, for example, the number 1 is an interior boundary of the union of $(0, 1)$ and $(1, 2)$. Regular open sets are those open sets that have no interior boundaries.

The domain of our first model consists of all non-empty, regular open sets of real numbers. The parthood relation is just the subset relation, as might be expected. Let Π be a non-empty set (or plurality) of non-empty, regular sets. Define $SUM(\Pi)$ to be the interior of the closure of the union of Π. This is the fusion relation. So to get the fusion of a set of regions in our model, first take the union of the sets, then the closure of the result, and then the interior of that. The result is, again, a regular, open set. And it is straightforward to verify that our Axiom 2, of fusion (or whole comprehension) is satisfied.

This model nicely recapitulates some of Aristotle's account of continuous objects. Consider, again, the intervals $(0, 1)$ and $(1, 2)$. Those are "contiguous", since there is nothing of the same kind in between them. Indeed, there are no members of our domain in between. The only thing "between" them is the real number 1, and $\{1\}$ is, of course, not a regular open set. These two intervals are also "continuous", in Aristotle's sense, since when we put them together—when we take their sum—we obtain the interval $(0, 2)$. That is, the "boundary" disappears and they become a single interval, a unity.

The proper definitions of the other primitives in our axiomatization are as straightforward as can be. Recall that $L(x, y)$ intuitively means that the region x is entirely to the left of y. Let X and Y be non-empty, regular open sets of real numbers. Then define X to be *LEFT* of Y just in case every member of X is smaller than every member of Y. It is trivial to verify that the relevant axioms are satisfied.

Recall that an "interval" is defined to be a connected, bounded region. In the model, the "intervals" are just the open intervals, (a, b), with $a < b$. And, of course, the Gunkiness Axiom 5 is also trivial:

$$(\forall x)(\exists j)(Int(j) \;\&\; j < x).$$

The notion of "left-end equivalence" is also straightforward. Two intervals are left-end-equivalent just in case they have the same left endpoint. And, of course, "congruence" of intervals is defined to be congruence of open intervals. Verifying the remaining axioms is also trivial.

Our second topological model is a sort of dual to the first, as its domain consists of certain *closed* sets of real numbers. It is, in one sense, a little more Aristotelian, since it does allow that intervals have endpoints—although these endpoints are not regions in the domain.

Say that a set S of real numbers is *regular closed* if S is identical to the closure of its interior. A real number r is an isolated point of a set B if $r \in B$, and there are numbers s, t such that $s < r < t$ and the open set $(s, t) - \{r\}$ is disjoint from B. So, for example,

1 is an isolated point of $\{1\}$, and also of the union of $\{1\}$ with $[2, 3]$. Regular closed sets have no isolated points.

The domain of our second model is the set of non-empty, regular closed sets of real numbers. As with our first model, the parthood relation is the subset relation. If Π is a non-empty set of non-empty regular closed sets, define the *SUM* of Π to be the closure of the union of Π. It is straightforward to verify that this *SUM* is regular closed, and that Axiom 2 of fusion is satisfied.

Perhaps this model better captures the Aristotelian notion of continuous objects. The sum of $[0, 1]$ and $[1, 2]$ is, of course, $[0, 2]$. Here the boundary point(s) of the contiguous intervals is "absorbed" into the sum.

With this model, we must be a little more careful when characterizing some of the relations in our theory, even the defined ones. Consider, for example, the two closed intervals $[0, 1]$ and $[1, 2]$, both of which are in our model. As sets, those are not disjoint, as they have a member, 1, in common. Recall, however, that in our definition of overlap:

$$x \circ y \leftrightarrow (\exists z)(z \leq x \,\&\, z \leq y),$$

the quantifier ranges over the *regions* in the model. So $[0, 1]$ does not overlap $[1, 2]$, since there is no member of the domain—no regular closed set—that is a part of (i.e. a subset of) both. These regions are indeed discrete. In our second model, "intervals"—bounded and connected regions—are closed intervals $[a, b]$.

Let A be the union of $[0, 1], [2, 3], [4, 5], \ldots$; and let B be the union of $[1, 2], [3, 4], [5, 6], \ldots$ Then A does not overlap B—they are discrete—even though their intersection, as sets, is infinite.

Our definition of "Left" is similarly nuanced. If A and B are regular closed sets, then define A to be *LEFT* of B just in case for every r in A and every s in B, either $r < s$, or both $r = s$ and r is a boundary of both. So $[0, 1]$ is to the left of $[1, 2]$. It is trivial to verify that the axioms of our theory are all satisfied in this second model.

This second model does have some rather strange or, at least intuitively unAristotelian regions. Consider the union of the closed intervals $[.1, .9]$, $[.01, .09]$, $[.001, .009], \ldots$ and $\{0\}$. This set is regular closed, and so is a region in our model. However, it has an actual infinity of discrete parts and a sort of loose point $\{0\}$ at its left end.

2.4 "Points" in *G*

Both of these topological models are defined in ordinary, Dedekind–Cantor space. Notice that in the first model, the intervals do not include their endpoints; while in the second, they do. We have been insisting that within the theory itself, there is no sense to be made of a given interval, say, being "open" or "closed" or "half-open". Let us sharpen that claim.

Clearly, our space *G* does not contain points as mereological parts. The gunkiness axiom takes care of that. Suppose, however, that we try to countenance the idea that, in some sense or other, a given interval is open, closed, or half-open.[16] Start with an arbitrary interval i. Since we have not stipulated anything either way, then, one would think it is left undecided whether, say, the right end of i is open or closed (i.e. whether it contains its right boundary, in some non-mereological sense of "contains").

For each natural number n, let $2^{-n}i$ be the result of bisecting i n times, taking the left bisect each time. In particular, $2^{-0}i$ is i, and, for each natural number n, $2^{-(n+1)}i$ is the left half of $2^{-n}i$. Define a *rightward Zeno sequence* of intervals as follows: Let b_1 be the left bisect of i; and, for each natural number n. Let b_{n+1} be the result of adjoining an interval congruent to $2^{-(n+1)}i$ immediately to the right of b_n. Now let i_1 be the fusion of the members of this sequence.

Continuing our assumption that it is somehow acceptable to talk about open/closed intervals, i_1 would be open on the right, at least intuitively. The interval i_1 does not include its right endpoint, since we never reach that point in the Zeno sequence, and so it would not be included in the fusion.

Now let j be any interval congruent with i. Define a sequence of intervals as follows: Let c_1 be the *right* bisect of j; and, for each natural number n, let c_{n+1} be the result of adjoining an interval congruent to $2^{-(n+1)}i$ immediately to the *left* of c_n. In effect, this is a Zeno sequence that goes in the other direction (leftward). Let j' be the fusion of the members of this leftward Zeno sequence. So, intuitively, it does not include its left endpoint (as the sequence never "gets there").

Return to our given interval i. Let $2i$ be the result of adjoining an interval congruent with i immediately to the right. Let j'' be congruent to j' and right-end-equivalent with $2i$. Finally, let i_2 be $2i - j''$. Now, intuitively, one would think that i_2 is *closed* at the right end—it does contain its right endpoint (again, in some non-mereological sense of containment). The reason is that, as we saw, j' and thus j'' do not contain their left endpoints, and so that point would remain when j'' is subtracted from $2i$. In effect, the right endpoint never gets subtracted.

So we have three intervals, i, i_1, and i_2. Intuitively, it is undecided whether i contains its right endpoint (i.e. whether it is open or closed at that end), while it seems that i_1 does not include its right endpoint (i.e. it is open at that end), while i_2 does include that endpoint (i.e. it is closed at that end).

In contrast with these assumptions, it is straightforward to show that $i = i_1 = i_2$: they are one and the same interval. Suppose, for example, that $i_1 \neq i_2$. Clearly, $i_1 \leq i_2$. So $i_1 < i_2$. It follows that $i_2 - i_1$ exists.

So, by Axiom 5, there is an *interval* that is a (mereological) part of i_2, but not part of i_1. A contradiction.

[16] Aristotle himself sometimes poses, and tries to answer, questions that sound a bit like this (if one can pardon the anachronism). Suppose, for example, that someone is at rest and then starts moving. Aristotle wonders if there is a first instant (or "now") of motion or a last instant of rest. We show here that if space and time are modeled in our framework, those questions have no answers—there are no points or instants to consider.

So there just is no sense to be made of saying that some intervals are open and others closed. The theory can recognize no such distinction.

We can however, follow the lead of the recovery of the real numbers and define "points" in our system by using the resources of the set-theoretic meta-theory, or superstructure. It is a kind of Whiteheadian "extensive abstraction" ([128], Chapter 7). Define a non-empty set S of intervals to be *convergent* if

(i) if $i \in S$ and $j \in S$, then either $i \leq j$ or $j \leq i$, so the intervals in S are linearly ordered by parthood,
(ii) if $i \in S$, $j \in S$, and $i \neq j$, then neither is right-end or left-end-equivalent to each other, and
(iii) there is no region that is part of every member of S.

The idea is that the members of S hone in on a "place" in the gunky line G.

Let S and S' be convergent sets of intervals. Say that $S \sim S'$ if for every $i \in S$ and $i' \in S'$, i overlaps i'. It is straightforward to verify that this is an equivalence relation.

So define a *"point"* (in scare quotes) to be an equivalence class on this relation. As usual, we will identify a given "point" with a representative member of the equivalence class.

Say that "point" p is *less than* "point" q just in case some interval in p is Left of some interval in q. In light of the above isomorphism with the real numbers, it is straightforward to show that these "points" are order-isomorphic to the real numbers, under less-than. They are indeed a Dedekind–Cantor continuum.

Finally, let p be a "point" and m a region in G. We can say what it is for p to be *interior* to m. This is just for some interval in p to be a proper part of m. Similarly, we can say what it is for p to be exterior to m. That is for some interval in p to be discrete from m. And we can say what it is for p to be a *boundary* of m. That is just for p to be neither interior nor exterior to m, or, in other words, for every interval in p to overlap m and for no interval in p to be a part of m.

However, as above, there is no sense in which any of these "points" are parts of any region. These "points" just are certain sets of sets of intervals.

2.5 Brief Philosophical Interlude

Objection: If your development really is coherent, it would indeed reinforce that "non-punctiform" and "indecomposable" are very different attributes as applied to continua.[17] *But the connection between them may be greater than the above seems to imply. Indeed, one may question whether it really makes sense to say of G that it has any proper parts*

[17] In the opposite direction, the intuitionistic continuum is "indecomposable" but, in some sense of 'point', is entirely composed of points. Although not all pairs of reals are orderable (so that "exact location" cannot be attributed), still they are specfied with infinite precision relative to the everywhere dense rationals.

at all! Let's take your interval i: You suppose that it makes sense to speak of i and its "negate", Neg(i) = G − i. But consider the question of the two places where i "meets" Neg(i): Do they touch there? Surely there must be a point at each of those places, either as part of i itself or of Neg(i). But of course, as your system allows, i can be shifted anywhere along G so that any place can serve as a boundary of an interval, in which case G is composed of points after all.

Reply: It is indeed part of the standard punctiform conception that boundaries of (bounded) intervals must exist, and then they must be part of one or the other of an interval and its complement. But the above theory of G makes sense on the basis of a *different* conception. According to it, let's consider your question, whether *i* and Neg(*i*) "touch". Well, we don't make sense of that, except to say that, if you mean, "Is there anything in between *i* and Neg(*i*)?", the answer is clearly no; and this is perfectly compatible with the two parts not overlapping. The regions are adjacent, in our technical sense.

Note that a very similar thing happens in the case of Dedekind cuts in the rationals used to define irrational numbers: the lower and upper sections are disjoint (the set-theoretic analogue of our "discrete"), yet there is no "boundary" in the sense of a lub of the lower section or a glb of the upper, until *by fiat* one is introduced by defining it to be the cut (or a section thereof) itself!

We all know and love this move by Dedekind, and we know how well it works. But that move does not in itself establish that irrational points are to be found literally among the rationals. One could instead treat the irrationals as superstructure over the field of rationals. And even the rationals could be treated as superstructure over the integers. (And, of course, further reductive moves are possible.) Indeed, we suggest that all reference to points can be coherently conceived as reference to superstructure constructed over a space of regions. And it is simply not a move that we are forced to make, either on conceptual or on practical grounds. Instead, we carry out a Dedekind style construction as a *superstructure* over G, preserving all the advantages of classical analysis while resisting the attribution of a point-ontology.

Objection: The analogy with Dedekind cuts (for irrationals) is flawed: Your G is supposed to model a continuum, so the analogue of a division of G into two discrete yet adjacent parts would be a Dedekind cut in the reals, not in the rationals. But the Dedekind continuum is complete, so there are no cuts in the reals without the corresponding real belonging to one segment or the other. This shows, does it not, that e.g. your Bisection axiom really makes no sense unless a point is added at the place where the two parts meet.

Reply: The analogy is imperfect; still, imperfect analogies can be of heuristic value. But it doesn't follow that points need to be added "where [the] two parts meet", for, as we said above, "the place where they meet" is language belonging to the Dedekind–Cantor theories of a punctiform continuum, but it is foreign language that can't really be translated into the theory of G. Point-like "places" simply are not recognized; *that* is the point! Yes, they can be introduced as superstructure, as shown above, but that doesn't require revising the description of *G*, or thinking that "really" the points are "there" on *G*.

We can speak (as we have spoken) of two intervals being, say, "left-end-equivalent" without thereby implying that there is a special further entity called "the left end (as a point)" shared by both segments. Even in cases where abstraction based on an equivalence relation may be reasonable and useful, e.g. as when we speak of "income levels" based on persons or families "sharing similar enough incomes", that doesn't mean—at least, without much further argument—that we have to recognize "income levels" as entities or that it is somehow incoherent not to.[18]

2.6 Comparisons with Some Other Constructions

To be sure, we are not the first to attempt point-free constructions of various mathematical domains. Without attempting anything like complete coverage of alternatives already in the literature, we present some comparisons that we hope will be instructive.

As our title states, we are concentrating on point-free constructions of the (so-called) classical continuum, $\mathbb{R}$, and in Chapter 5 we develop a recovery of $\mathbb{R} \times \mathbb{R}$, which extends to higher dimensions, without primitives for any objects of lower dimension. In addition to forming the basis of classical functional analysis, these systems can be adapted to study geometric spaces of various sorts, including hyperbolic and spherical geometries, in Chapter 6. So we are clearly working in the area (category) of *metric spaces*, not purely topological spaces.

There have been a number of efforts to develop topological spaces based on axioms governing *regions*, rather than sets of points, going back at least to work of Karl Menger (e.g. [85]). Modern point-free topology has taken an algebraic turn, investigating Heyting lattices of open sets of a topological space, with important results in the spirit of Stone's representation theorem for Boolean lattices. For an overview, see Johnstone [67], and for key results, see Gierz et al. [48], especially ch. 5 (5.16, 271). For related approaches, see also Arsenijević and Kapetanović [8] and Arsenijević and Adžić [1], Roeper [98], and Gruszczyński and Pietruszczak [50].[19]

It is true that the classical Dedekind–Cantor continuum can be arrived at quite elegantly by means of the modern lattice-theoretic approach to point-free topology. The method involves defining points via a 1-1 correspondence between points of the underlying topological space and the maximal open sets of the Heyting algebra of open sets resulting from "subtracting singletons", identifying a "left half" and a "right half", and then defining one such "point" to be "left" of another if its left half is a subset

[18] The connection between this sort of abstraction and so-called Scottish neo-logicism is developed in [111].

[19] Bostock [24] is an attempt to present Whitehead and Russell's own point-free account of space and time.

(or part) of the other's. However, the axioms governing these regions are simply identified as those specifying a complete linear ordering. In effect, Dedekind completeness and, as a consequence, the Archimedean principle, are just *stipulated*. In contrast, as we have already shown in detail, we are able to derive suitable expressions of these as theorems in our regions-based system, without a detour through isomorphism with $\mathbb{R}$. We thus preserve, in our approach, one of the few lasting achievements of "logicism", viz. *deriving* the continuity principle in our logical framework, as did Dedekind in his, rather than simply postulating it as an extra axiom. Presumably a derivation like ours could also be carried out in the algebraic framework by translating the steps we have taken. But from an axiomatic standpoint, the algebraic approach does not provide an independent, equally advantageous system. This, of course, is not to deny that it affords other insights into the sources of continuity and related mathematical phenomena.

In any case, our own work only touches indirectly on these precedents. Focusing, then, on point-free geometries, there is the noteworthy reconstruction of three-dimensional Euclidean geometry by Tarski [119]. This explicitly uses (atomless) mereology, to which is added the single primitive '*sphere*'. We consider that approach more directly in Chapter 5.

A more recent development is the work of Arsenijević and Kapetanović [8] and Arsenijević and Adžić [1], which build axiom systems, respectively, for one- and two-dimensional regions-based and point-based continua, in infinitary languages $L\omega_1\omega_1$, and then show that regions-based and point-based theories (of the same dimension) are relatively interpretable in one another. A detailed comparison of these theories with those developed here would take us too far afield. Suffice it to say that, although the spirit and main philosophical conclusions of the two approaches are quite similar, the details of the theories developed therein are very different. To mention just three significant points of comparison: One reason given for the choice of infinitary language is that ontological commitments beyond those objects—regions or points—making up the continua themselves (such as sets and relations among regions or points) are thereby avoided. This is also achieved in the present approach so long as our second-order resources are framed in the language of plurals. Second, we believe our axioms systems and definitions are significantly more streamlined than those being cited, and they achieve greater unification through their novel derivations of Dedekind completeness and, autonomously, of the Archimedean properties. Third, for dimensions 2 and higher, our systems derive key geometric structures, both Euclidean and non-Euclidean.

Closer in content and method to the reconstructions of continua presented here is the recent work of Peter Roeper [99] on what he calls "the Aristotelian continuum". Like our system presented above, his characterizes a linear continuum on a point-free basis in terms of a structure of regions and intervals, and it is shown to be categorical and isomorphic to a classical, point-based continuum. The following are the main points of comparison:

(1) Roeper's axioms describe a continuum as a certain kind of regions-based topological space, connected, locally connected, second-countable, also linearly ordered, complete, and separable. Our axioms do this but also describe the metrical structure of an ordered field. Thus, we have *congruence* of intervals as one of our primitives, whereas Roeper's system omits this.

(2) As part of his logic, in place of mereology and a logic of plurals, Roeper uses a (first- and second-order) logic of mass terms. Thus, where we would express, e.g. "region r is entirely to the left of region s" as "any part of r is left of every part of s", Roeper's language expresses "r is everywhere left of s (everywhere)". Probably the two languages, as used in these respective reconstructions of continua, are inter-translatable, at least in the presence of other primitives and axioms governing them.

(3) Roeper draws heavily on regions-based topology, a theory with two non-logical primitives, a monadic predicate for *limited*, meaning "bounded", or "finite in extent", and a binary relation α *is connected with* β, written $\alpha \infty \beta$, intuitively meaning that either α and β overlap or they "abut one another". He presents thirteen axioms governing these, including *Coherence* (if $\gamma = \alpha \cup \beta$, then $\alpha \infty \beta$), *Continuity* (a topological version of a least-upper-bound or greatest-lower-bound principle), and *Countable Convex Cover* (akin to Separability). Axiom 6 is essentially our definition of *bounded*; Axiom 7 is a version of *continuity*; and Axiom 8 is a version of *separability*.

Roeper's "limited" is the same as our defined notion of "bounded", and "connected" is the disjunction of our adjacent and the mereological notion of overlap. His crucial properties of *coherence*, *continuity*, and *separability* are not taken as axioms of our system, but rather proved as theorems, as we have shown (above and in Appendix 2.3), where for the first two properties, we use the plurals comprehension scheme of mereology (describing minimal closures and certain fusions), and for the third we use the representation of rational-length intervals via the theorem of bisectability. Regarding the important Archimedean property, Roeper's system can derive this from its version of continuity or completeness (as standardly done in point-based frameworks, e.g. à la Dedekind), whereas we prove this directly from our comprehension axioms and Translation.

Indeed, we can claim more: all of Roeper's axioms are in fact derivable as theorems in our system, under the translation just given of his primitives, *limited* and *connected*. Thus we have a nice unification of systems, our geometry-*cum*-classical analysis and Roeper's regions-based topology-*cum*-ordered continuum. In sum, "honest toil" has paid off.

Appendix 2.1 Proof that the interval structure of G is Dedekind complete (satisfies the LUB principle)

We use j, k, m as variables for intervals (recall that intervals are bounded, connected regions), X, Y, etc. for sets or pluralities of intervals, and 'R' ('L') for right (left).

Some definitions:

Definition 1. *Let X be a set of intervals. Say X contains a left-most interval iff there is an interval k of X such that, for any interval m of X that is not left-end-equivalent with k, a part k′ of k is left of m.*

Definition 2. *Let X be a set of intervals, and let $Y \subseteq X$; then we say that Y extends to the right exactly as far as X iff no part of any of the Y is right of all the X, and no part of any of the X is to right of all the Y.*

Definition 3. *Let X be a set of intervals. We say that X is R-bounded iff* $\exists j \forall k[k \in X \rightarrow \exists j'(j' \leq j \;\&\; R(j', k))]$.

For the next definition, we need the following construction: Given a set or plurality X of intervals, let Y be any subset of X such that Y contains a left-most interval but extends to the right exactly as far as X. Now let Y' be the result of adding to Y any interval between any two of the Y intervals (calling such Y' "the gap-filled extension of Y"); and let F be the fusion of Y'. By an axiom of mereology on forming fusions, F is the meet of all individuals H which overlap all individuals overlapping some interval of Y'. We say that F is *the least* such H. F then overlaps exactly all individuals which overlap some interval of Y', by definition *the fusion of* Y'.

Definition 4. *Let X be an R-bounded set of intervals. Then we say that j R-bounds X iff j witnesses that X is R-bounded (Df. 3), or for some $Y \subseteq X$ with a left-most interval and extending to the right exactly as far as X, j is R-end-equivalent to F, the fusion of Y', the gap-filled extension of Y (as above).*

Note that, by construction, F is an interval: As the fusion of Y', F is connected. Furthermore, Y has a left-most interval so F is left-bounded; and by the hypothesis of the definition, X and hence Y and F are right-bounded. Thus, the condition "j is R-end-equivalent to F" makes sense. (End-equivalences are relations of *intervals.*)

Definition 5. *j is a least R-bound of X iff j R-bounds X* $\&\, \forall k[k$ *R-bounds X & k not R-end-equivalent to* $j \rightarrow \exists k'(k' \leq k \;\&\; R(k', j))]$.

Theorem 6. *(LUB) Let X be an R-bounded set of intervals; then there exists a least R-bound of X, unique up to R-end-equivalence.*

Proof. (Sketch): Step 1: Carry out the construction of F as described above after Df. 3. As noted after Definition 4, by construction F is an interval.

Step 2: Claim: F is a least R-bound of X.

Proof of Claim: 1) F R-bounds X, for, by the construction of F, F extends at least as far right as any k of X. 2) F is a *least* R-bound of X: Let F' be any R-bound of X not R-end-equivalent to F. Then by the construction of F as the least region satisfying the "gap-filling" condition, we must have $\exists m[m \leq F' \;\&\; R(m, F)]$, for otherwise, since F' is not R-end-equivalent to F, there would be some $n \leq F$ with $R(n, F')$. But by the construction of F, some part of X is also right of F', contradicting the hypothesis that F' is a R-bound of X.

That the uniqueness clause is satisfied follows from axioms of mereology. ■

The argument that there is a greatest lower-bound of any L-bounded set of intervals is exactly analogous.

Appendix 2.2 Doing without "Left" and "Right"

Recall our "geometric" two-place predicate, $L(x, y)$, meaning that "region x is (entirely) to the left of region y". One might object that this is somehow extraneous to the continuum. Intuitively, a continuum does not come with designated "directions". We submit that the objection is misguided. A one-dimensional continuum, like the present G, does have two directions, and we are just labeling one of them as "left". It does not matter which is which.[20]

Nevertheless, we show here how to get by without this primitive. As it happens, the important notion is that of one region being *between* two others. *This* relation is, intuitively, intrinsic to the continuum. In the above, betweenness is defined; instead, we can take it as a primitive.

So $Betw(x, y, z)$ is to mean that y is entirely between x and z. For axioms, we adopt the following:

$$Betw\,(x, y, z) \rightarrow (x|y \;\&\; x|z \;\&\; y|z)$$

If y is between x and z, then they are pairwise disjoint.

$$Betw\,(x, y, z) \rightarrow Betw\,(z, y, x)$$

$$Betw\,(a, b, c) \leftrightarrow \forall x, y, z[(x \leq a \;\&\; y \leq b \;\&\; z \leq c) \rightarrow Betw\,(x, y, z)].$$

Recall that in the above treatment, "connected", "bounded", "interval", and "adjacent" are all defined in terms of *Betw*. Those definitions stand.

Our Dichotomy Axiom 4 is stated in terms of "left" and "right". In its place, we adopt the above Theorem 2, linearity, as an axiom:

Let x, y, z be any three pairwise discrete intervals; then exactly one of x, y, z is between the other two.

The most significant and, admittedly, the most awkward change is a replacement for "left-end-equivalent", and, more important, the Translation Axiom 6.

Let c, j, and k be intervals. Say that j and k are *c-end-equivalent* just in case (i) c is adjacent to j, (ii) c is adjacent to k, and (iii) $j \circ k$. The idea, intuitively, is that j and k are on the same side of c.

Proofs of the following are tedious, but straightforward:

Lemma: For a fixed c, the relation of c-end-equivalence is symmetric and transitive.

Lemma: Suppose that j is c-end-equivalent to k. Let d be any interval such that d is adjacent to j and $d \circ c$. Then j is d-end-equivalent to k.

The key here is that it does not matter which interval we pick for the "c" role.

Now we come to the replacement for the Translation Axiom 6:

Let c, i, and j be intervals, such that c is adjacent to i. Then there is an interval j' such that j' is congruent to j and j' is c-end-equivalent to i.

The rest of the development goes through, with straightforward adjustments.

[20] Thanks to Kit Fine and Øystein Linnebo for pressing the objection, and to Linnebo and Salvatore Florio for suggesting the present resolution. The situation is analogous to that in pure complex analysis, which posits two square roots for -1. We just use 'i' for one of them—not caring which. It follows that $-i$ is the other. See [109].

Appendix 2.3 Non-set-theoretic versions

There were two main places in the above reconstructions where set theory seemed to play an indispensable role: (1) in the use of Cauchy sequences of binary intervals and of equivalence classes of sets of regions in first the recovery of "points"; and (2) in the use of isomorphisms between our gunky line G and the real numbers $\mathbb{R}$. In the latter case, appeal to set theory is, in a sense, not really problematic with regard to the autonomy of the interval structure of G, since, after all, we're reasoning about the relation of that structure to that of the (so-called) classical continuum, which is fundamentally set-theoretic. The first case, however, is problematic, as the autonomy of G is threatened by reliance on enough set theory to build up $\mathbb{R}$, with its ontology of points, etc. Is there an alternative way of introducing interval sequences that improves on this?

Indeed there is. We will stick here to the use of Cauchy sequences of binary intervals. The other recovery of "points" requires no new conceptual resources, except perhaps to do the construction with sequences instead of sets.

What is needed is to expand the theory of G by specifying that, additional to G but entirely discrete from it, there is a denumerable infinity of mereological atoms that form a natural-numbers structure ("N-structure"). That is readily expressed with mereology and plural quantification. (Cf. e.g. Hellman [57].) Now the role of ordered pairs of the form $\langle n, j\rangle$, where n is an atom of the N-structure and j is an binary rational interval of G, can be served by the (unordered) mereological sum, $n + j$.[21] Plurally quantifying over such things then enables *plurally quantifying* over binary rational interval sequences. Those that satisfy the Cauchy convergence condition serve as our reals. The upshot is that, now, we bypass set theory entirely in the development of our G-interval structure, *even to the extent of being able to carry out a great deal of classical analysis without even recognizing real numbers as objects!* Instead of speaking of reals as individual objects, we quantify plurally over the increasing binary rational Cauchy interval sequences corresponding to them (as their limit points, on standard classical theory). This fulfills at least the non-punctiform aspect of the (semi-)Aristotelian conception of contiua in a clearly non-parasitic way. Points aren't introduced at all; rather, their mathematical roles are performed by surrogates within the thoroughly non-punctiform framework of G.

But, in order to make good on this last claim, it needs to be checked that at least a good portion of classical analysis can indeed be constructed within the G-framework (with plural quantification, but no set theory, as just described). In particular, how are we to reconstruct quantification over functions from reals to reals if we can only plurally quantify over Cauchy interval sequences? This is no problem so long as our functions are continuous (or at least continuous on a co-countable domain of reals). The key here is that continuous functions are determined by their behavior at rational arguments. Now binary rationals are available directly as intervals. Suppose we have a sequence of all these representing rationals in the domain of definition of f, i.e. we refer plurally to countably many individuals of the form $\langle n, j\rangle$, as defined

[21] Although the intervals of G already allow encoding of natural numbers and the arithmetic operations, representation of sequences of intervals and other functions within a nominalist framework becomes cumbersome at best, since a stock of intervals would be serving two roles at once, that of natural numbers and that of items of sequences. Such problems are bypassed by positing an atomic N-structure discrete from G.

above. A continuous function f assigns each of these a real value, represented over G as a Cauchy interval sequence, $f(\langle n,j\rangle) = f_1^{n+j}, f_2^{n+j}, \ldots$ Now we would like to code all these (countably many) Cauchy interval sequences as a single interval sequence (which itself, of course, need not be convergent—it merely serves to encode the behavior of f at rational arguments, enumerated in an arbitrary fashion). One convenient way is to work with just the atoms n of the $\langle n,j\rangle$ serving in the enumeration of rational intervals, coding the (set-theoretic) ordered-pair $\langle n, f(\langle n,j\rangle)\rangle$ as the sequence $n+f_1^{n+j}$, $n+f_2^{n+j}, \ldots$ Then the restriction $f \restriction \mathbb{Q}$ of f to rationals can be coded up in a single sequence by a dovetailing construction on the countably many sequences of this latter form.[22] In this way, we bring continuous functions within the purview of plural quantification of the language of the G-interval structure.

With one more reductive step, we can improve further on this to bring also the isomorphism recovery theorems within the purview of the theory of G. Instead of recovering real numbers as Cauchy interval sequences, we can avail ourselves of the second relative interpretation of $\mathbb{R}$ over G referred to above, toward the end of section 2.2, viz. representing real numbers as certain intervals of G, viz. fusions of the right-(or left-) increasing Cauchy interval sequences introduced above. Here is how this helps in representing the isomorphisms of the recovery theorems without set-theoretic machinery.

As in the non-set-theoretic reconstruction just sketched, we reconstruct interval sequences via an atomic N-structure. But now instead of treating quantification over reals as plural quantification over the wholes, $n+j$, coding the (canonical) Cauchy interval sequences defining them, instead we quantify singularly over the fusions of the intervals making up such a sequence, which, as noted, are themselves intervals of G.

Before we can introduce functions from such "real intervals" of G (as we may now call them) to reals of the classical continuum, we need (or at least prefer) to identify a suitable target structure that is itself not essentially set-theoretic. Here we can appeal to arithmetic methods applied to our given N-structure. Coding signed integers $+m$ ($-m$) conventionally as certain pairs of naturals, e.g. as $2^1 \cdot 3^m$ ($2^2 \cdot 3^m$), and then rationals as (reduced) pairs of signed integers, we then have canonical Cauchy sequences $\langle q_j\rangle$ of rationals coded as number-theoretic ordered pairs of the form $\langle j, \overline{q}_j\rangle$, where $\overline{q}_j$ is the number coding the rational q_j.[23] Any two distinct such sequences differ at some place; thus, the fusion of any such sequence is just a uniquely determined whole of atoms of our original N-structure, with different sequences yielding different fusions. These fusions then can serve as our classical reals (which we call "reals over $\mathbb{N}$"). It is tedious, but routine, to define a linear ordering and the field operations on these.

We still need a way of representing or coding mappings between real-intervals of G and fusions over the N-structure serving as classical reals. First, how can we code an ordered pair of the form $\langle k, \sigma\rangle$, where k is a real-interval of G and σ is a (nominalistic) sum of "naturals" (atoms of the N-structure) coding a canonical Cauchy sequence of rationals representing a real? Well, if we stipulate that the mapping goes in the direction $k \rightarrowtail \sigma$, we can make do with an *un*ordered pair. But that can simply be taken as the fusion of k and σ, as these are always non-overlapping, as they are from different structures, one atomless, the other atomic. Finally, by quantifying

22 This is modeled on the way countably many real numbers can be coded as a single real number.
23 Here, of course, j is a numerical index referring to an atom of our N-structure, not an interval of G.

plurally over fusions of this form $(k + \sigma)$, we achieve the effect of quantifying over mappings or functions from our G-structure to the reals over $\mathbb{N}$. In this manner, we can reconstruct the isomorphisms of the recovery theorems and prove their required properties within our theory of G together with an atomic N-structure, without ever using set-membership or quantifying over sets.[24]

[24] We might add that the entire development above can be set in the modal-structural framework of Hellman [56], as improved in [57] and subsequent presentations. So we only have to assume the *logical possibility* of a gunky interval structure. Extensions of G by further structures, such as an atomic N-structure as above, are taken as *further* possibilities, relative to any given hypothetical G-structure, so that the "merely possible" status of real numbers as points is still recognized even within a thoroughgoing modal-structural treatment.

We do not claim that this captures what Aristotle meant by "potentially infinite". This is attempted, in part, in Chapter 3.

3

Aristotelian and Predicative Continua

With Øystein Linnebo

3.1 Introduction

The previous chapter develops a "regions-based" one-dimensional continuum, following the Aristotelian theme that continua are not composed of points. Each part of our "gunky" space—each region—has a proper part. Although we show how "points" can be defined on the structure in terms of sets or sequences of regions, along lines similar to Alfred N. Whitehead's "extensive abstraction" ([128], 7), there is no sense in which "points", so construed, are in, or are part of, regions. Indeed, "points", so construed, just *are* set-theoretic constructions on regions. We show that these "points" are isomorphic to the real numbers, and thus to the so-called "classical" point-set continuum of Dedekind and Cantor.

As noted, the theory does not follow Aristotle in another, important aspect. Recall that Aristotle, along with almost every major mathematician and philosopher before the nineteenth century, rejected the very notion of the actual infinite. They argue that the only sensible notion is that of potential infinity (at least for scientific or otherwise non-theological purposes). In *Physics* 3.6 (206a27–9), Aristotle wrote, "For generally, the infinite has this mode of existence: one thing is always being taken after another, and each thing that is taken is always finite, but always different". As Richard Sorabji [116] (322–3) puts it, for Aristotle, "infinity is an extended finitude" (see also [70], [71]). Aristotle, and just about every ancient, medieval, and early modern mathematician, recognize the existence of certain *procedures* that can be iterated indefinitely, without limit. Examples are the bisection and the extension of line segments. What they reject are what would be the end results of applying these procedures infinitely often: self-standing points or infinitely long regions. They would also object to thinking of a sequence as itself an actually infinite entity.

In contrast to that, the theory presented in the previous chapter makes essential use of the actual infinite, much in line with contemporary mathematical practice. We thus dub that theory *semi-Aristotelian*. One goal of the present chapter is to provide a more

Aristotelian account of continuity, sticking, again, to one dimension. The first version is obtained by modifying the theory from the previous chapter to bring it in line with the rejection of actual infinity. It turns out that in the semi-Aristotelian theory, the use of actual infinity allows one to prove several theorems that are, in fact, legitimate for the Aristotelian. So, in the Aristotelian theory, analogues of several key theorems have to be added in by hand, as new axioms. This at least suggests the fruitfulness of the notion of actual infinity.

Our second Aristotelian theory attempts to capture the "potential" nature of Aristotelian continua in a modal language. Building on some results of [79], [81], we show that the modal and non-modal theories are equivalent. We suggest, however, that the modal framework is the more natural, in that it better expresses the "potential" nature of the various procedures. Moreover, the modal language has the expressive resources to formulate more of the Aristotelian themes.

We then present a theory that, intuitively, lies between the Aristotelian and the semi-Aristotelian ones. It allows the use of (countably) infinite sets (or properties or pluralities) only if they have *predicative* definitions. Many of the usual issues concerning predicative analysis have direct analogues here (see [42], [43]).

In the interests of readability, we begin with brief statements of the axioms and results of the semi-Aristotelian theory of the previous chapter. Along the way, we point out the places where actual infinity is invoked.

3.2 Recapitulation of the "Semi-Aristotelian" Continuum (with Emphasis on "Semi")

The formal background is classical, first-order logic with identity supplemented with a standard axiom system for second-order logic (or a logic of plural quantification, with an unrestricted comprehension axiom for plurals). The Aristotelian systems developed in subsequent sections do not invoke these higher-order resources, and the predicative system restricts them accordingly.

1a. Axioms on $x \leq y$ ("x is part of y"): reflexive, antisymmetric, transitive.

As is standard, we define a binary relation called "overlaps": $x \circ y \Leftrightarrow^{df} \exists z(z \leq x \,\&\, z \leq y)$. And we write $x|y$ for $\neg x \circ y$, pronounced "x is discrete from y".

1b. Axiom on $\leq$ and $\circ$: $x \leq y \leftrightarrow \forall z[z \circ x \rightarrow z \circ y]$.

This implies an extensionality principle, yielding so-called classical mereology:

Theorem 1. $x = y \leftrightarrow \forall z[z \circ x \leftrightarrow z \circ y]$.

2. Axiom of Fusion or Whole Comprehension: $\exists u \Phi(u) \rightarrow [\exists x \forall y\{y \circ x \leftrightarrow \exists z(\Phi(z) \,\&\, z \circ y)\}]$, where Φ is a predicate of the second-order language (or language of plurals) lacking free x.

This is the axiom that allows the play with infinity, so to speak. In the semi-Aristotelian theory, there are no restrictions on the number of regions that can be "fused". In some cases, we take fusions of regions without bothering to check how many regions are so fused. In other cases, we explicitly fuse infinitely many regions. We take it that the "infinities" invoked by this axiom are "actual", since we conclude that there is a *single* region that is the fusion of *all* of the given regions.

We write $x + y$ for the mereological sum or fusion of x and y. If $x \circ y$, then we write $x \wedge y$ for the *meet* of x and y; it is the fusion of *all* regions that are part of both x and y. So $\forall z[z \leq x \wedge y \leftrightarrow z \leq x \,\&\, z \leq y]$ (and if x and y have no common part, then $x \wedge y$ is undefined).

Here, then, is the first place where we take the fusion of some regions without bothering to make sure that there are only finitely many such regions.

Similarly, if $\exists z(z \circ x \,\&\, \neg(z \circ y))$, then we let $x - y$ be the sum of all regions z that are part of x but discrete from y (and if there is no such z, then $x - y$ is undefined). So $\forall z(z \leq x - y \leftrightarrow (z \leq x \,\&\, \neg(z \circ y)))$. Again, we do not check to see how many regions are so fused.

We let G be the fusion of *all* regions. It is the entire line, as a single region. Since, as it happens, there are infinitely many regions, this, too is not acceptable to an Aristotelian.

We introduce a geometric primitive, $L(x, y)$, to mean "x is (entirely) to the left of y". The axioms for L specify that it is irreflexive, asymmetric, and transitive. And we define '$R(x, y)$', "x is (entirely) to the right of y", as $L(y, x)$.

This allows us to define *betweenness*: $Betw\,(x, y, z)$ for "y is (entirely) between x and z":

$$Betw\,(x, y, z) \Leftrightarrow^{df} [L(x, y) \,\&\, R(z, y)] \vee [R(x, y) \,\&\, L(z, y)]$$

It follows that $Betw\,(x, y, z) \leftrightarrow Betw\,(z, y, x)$.

$L(x, y)$ obeys the following axioms:

3a. $L(x, y) \vee R(x, y) \rightarrow x|y$.
3b. $L(x, y) \leftrightarrow \forall z, u[\, z \leq x \,\&\, u \leq y \rightarrow L(z, u)]$.

The following can now be inferred:

$Betw\,(x, y, z) \rightarrow x|y \,\&\, y|z \,\&\, x|z$, and

$Betw\,(x, y, z) \,\&\, Betw\,(u, x, z) \rightarrow Betw\,(u, y, z)$.

We next define the notion of a *connected* region, one that has no gaps:

$$Conn(x) \Leftrightarrow^{df} \forall y, z, u[z, u \leq x \,\&\, Betw\,(z, y, u) \rightarrow y \leq x]. \qquad \text{(Df } Conn\text{)}$$

We define what it means for a region to be *bounded* (on both sides):

$$Bounded(p) \Leftrightarrow^{df} \exists x, y\, Betw\,(x, p, y). \qquad \text{(Df } Bounded\text{)}$$

We call bounded, connected regions "*intervals*" and write '$Int(j)$' etc., when needed. Using L, we can impose a condition of dichotomy for discrete intervals:

4. Dichotomy axiom: $\forall i, j[i, j$ are two discrete intervals $\rightarrow (L(i, j) \vee L(j, i))]$.

Next we prove a linearity condition among intervals:

Theorem 2. *(Linearity): Let x, y, z be any three pairwise discrete intervals; then exactly one of x, y, z is between the other two.*

To guarantee that arbitrarily small intervals exist everywhere, we adopt the following "gunkiness" axiom:

5. $\forall x \exists j[Int(j) \,\&\, j < x]$.

An important relation of two intervals is "adjacency", which is defined as follows:

$$Adj(j,k) \Leftrightarrow^{df} j|k \,\&\, \nexists m[Betw\,(j,m,k)]. \qquad \text{(Df } Adjacent\text{)}$$

Intuitively, adjacent intervals touch, but do not overlap.

The following equivalence relation is useful: "*j and k are left-end-equivalent*" just in case

$$\exists p[p \leq j \,\&\, p \leq k \,\&\, \nexists q(\{q \leq j \vee q \leq k\} \,\&\, L(q,p))].$$

"*Right-end-equivalent*" is defined analogously.

Our final geometric primitive is *congruence*, a binary relation among intervals. We adopt the usual first-order axioms specifying that *Cong* is an equivalence relation. We will sometimes write this as $|i| = |j|$. Similarly, for intervals i, j, we can define, contextually, $|i| < |j|$ as meaning: $\exists j'[j'$ an interval $\& \, j' > j \,\&\, \text{Cong}(i,j')]$; and we may write $|i| > |j|$ as equivalent to $|j| > |i|$. Say that $|i| \leq |j|$ just in case either $|i| > |j|$ or $|i| = |j|$.

The next axiom is crucial to our characterization:

6. **Translation axiom:** Given any two intervals, i and j, each is congruent both to a unique left-end-equivalent and to a unique right-end-equivalent of the other.

Lemma 1. *Given any two intervals i and j such that $\neg Cong(i,j)$, either there exists an interval $i' < j$ with $Cong(i,i')$; or there exists i' with $j < i'$ with $Cong(i,i')$.*

Theorem 3. *(Trichotomy) For any two intervals, i, j, either $|i| = |j|$ or $|i| < |j|$ or $|i| > |j|$.*

Our final axiom is that congruence respects nominalistic summation of adjacent intervals:

7. **Additivity:** Given intervals i, j, i', j' such that $Adj(i,j)$, $Adj(i',j')$, $Cong(i,i')$, $Cong(j,j')$, then $Cong(k,k')$, where $k = i + j$ and $k' = i' + j'$.

We now turn to the matter of the bi-infinitude of *G*. In fact, our axioms already guarantee this, as we can prove:

Theorem 4. *(Bi-Infinity of G) Let any interval i be given; then there exist exactly two intervals, j, k, such that $Cong(i,j)$ & $Cong(i,k)$ & $Adj(i,j)$ & $Adj(i,k)$ & one of j, k is left of i and the other is right of i.*

The proof of this also involves taking the fusion of some intervals and, here too, there is no reason to think, in advance, that the number of intervals so fused is finite.

Call an interval l an (*immediate*) *biextension* of interval i—$BiExt(l, i)$, or $biext(i) = l$—just in case $l = j + i + k$, where j, i, k behave as in the Bi-Infinity theorem.

Lemma 2. *Let i and j be intervals such that $i < j$; then $\neg Cong(i, j)$.*

One central result is the Archimedean property that, in effect, our entire space G is *exhausted* by iterating the process of flanking a given interval by two congruent ones as in Bi-Infinity.

Let X be any class (or plurality) of intervals such that an arbitrary but fixed interval $i \leq G$ is one of the X and such that if $k = biext(j)$ for j any of the intervals of X, then k is also in X. Call such X a "closure of i under *biext*".

Lemma 3. By axiom 2, there is an individual which is the common part of the fusions of each class X which is a closure of i under *biext*, which we call their *meet* or *the minimal closure i^* of i under biext*. (Since i is stipulated to belong to any such X, the meet is non-null, as required in mereology.)

Here we apply Axiom 2, of unrestricted fusions, this time to an *explicitly defined* infinite set (or plurality) of intervals. Moreover, the definition of this "minimal closure" is impredicative.

This is sufficient to establish an Archimedean property:

Theorem 5. *(Characterization of G): Let G be the fusion of the objects in the range of the quantifiers of our axioms; then $G = i^*$, the fusion of the minimal closure of i under biext.*

We then establish that any interval has a unique bisection:

Theorem 6. *(Existence and uniqueness of bisections): Given any interval i, there exist intervals j, k such that $j < i \,\&\, k < i \,\&\, j|k \,\&\, j + k = i \,\&\, Cong\,(j, k)$; and j, k are unique with these properties.*

Here, again, we take the fusion of some intervals, without checking to see how many intervals are fused.

We define a sequence $\langle j_i \rangle$ of intervals increasing to the right (or left, for negative reals) to be *Cauchy* just in case, for any interval, ε, there exists a natural number N such that for any natural numbers $m > k > N$, $j_m - j_k$ is an interval, $R(j_m - j_k, j_k)$, and $|j_m - j_k| < |\varepsilon|$. The fusion of the members of such a sequence gives us an interval that can play the role of a given real number. But, of course, these "Cauchy sequences" of intervals are themselves infinite and so here, too, we take the fusion of an infinite set of regions. We show how to embed the real numbers into G, given an arbitrary interval i to serve as a unit.

3.3 Basic Aristotelian Theory

The above semi-Aristotelian treatment invokes infinity in two related ways, neither of which is acceptable to an Aristotelian. First, there are regions, such as the entire Gunky line G, that are infinitely long, so to speak. Specifically, G is infinitely long relative to any interval. This difference with Aristotle (and Euclid) is, we think, fairly minor. One can just restrict the focus to regions that are "bounded", in the above sense. The Archimedean property entails that bounded regions are all finite, relative to each other (and thus to each interval).

The second and more substantial difference with Aristotle is that in the proofs of some of our theorems, we take fusions of the regions in infinite sets (or the regions in infinite pluralities, or the regions that bear properties with infinite extensions). For an Aristotelian, there is no conceptual problem with there being a *potential* infinity of regions. Ancient geometry establishes that there are certain procedures, like taking bisections and biextensions, and these procedures can be applied over and over, without limit. But, it seems to us, one cannot take the fusion of infinitely many such regions—to produce a *single* region—unless we take the infinity to be "actual".

It turns out that in the Aristotelian theories, developed below, analogues of several of the key *theorems* above have to be added in by hand, as new *axioms*. This at least suggests the fruitfulness of the notion of infinity. Recall that our play with infinity is sanctioned by the above Axiom 2, a principle of fusions or whole comprehension:

$$\exists u\Phi(u) \rightarrow [\exists x\forall y\{y \circ x \leftrightarrow \exists z(\Phi(z) \,\&\, z \circ y)\}],$$

where Φ is a predicate of the second-order language (or language of plurals) lacking free x.

Recall that there are a few places where we *explicitly* take the fusion of an infinite set (or plurality) of regions. A trivial example is the result that there is a "universal region" G, the fusion of all regions. A more substantial use is in the proof of the Archimedean property, in Theorem 5. That involves the minimal closure of an arbitrary interval i under biextension. This minimal closure, of course, is an infinite set (or plurality or property). We then take the fusion of the regions in this minimal closure and show that it is identical to the entire space G. In recovering counterparts of the real numbers, we likewise take fusions of infinite sets/pluralities, namely the members of certain "Cauchy sequences" of intervals. In light of the other axioms and theorems, there are infinitely many such regions in all of the non-trivial Cauchy sequences.

Also, as a reminder, there are places where we invoke the axiom of fusions without any guarantee that the fused set (or plurality) is finite. For example, if $x \circ y$, we define the *meet* $x \wedge y$ of x and y to be the fusion of *all* regions that are part of both x and y.

One natural Aristotelian "fix" to our system would be to replace the above Axiom 2, of unrestricted fusions, with an axiom asserting the existence of fusions of finite sets.[1] For this, of course, it is sufficient to state that any two regions have a fusion:

$$\forall u \forall v \exists x \forall y [y \circ x \leftrightarrow (y \circ u) \vee (y \circ v)]. \qquad \text{(FINFUS)}$$

A referee of a previous version of this chapter argues that the present theory "is not a very Aristotelian enterprise". In particular, the FINFUS axiom "allows us to make a region out of regions that are (to put it colloquially) nowhere near one another". This referee claims that scattered regions like this are not objects in the Aristotelian world: "This is important, because Aristotle is committed to the view that mathematical concepts are presented in sense experience".

Although we are not scholars of this period, it does seem to us that some scattered objects, like pairs of shoes or a given armada, *are* presented in sense experience. Moreover, Aristotle does seem to allow geometric objects that are *not* abstracted from things presented in sense experience. Consider, for example, the sphere that contains the fixed stars or perhaps the circles that are the orbits of the various heavenly bodies.

In any case, it is consistent with our presentation that the only *metaphysically basic* regions are intervals. One can take the talk of scattered regions as just a manner-of-speaking, a way to refer to intervals, in the plural (perhaps in line with the metaphysical principle expressed in note 1).[2] To be sure, we *define* "interval" in terms of "region" generally. Indeed, our definition gives the central properties of intervals, namely that they are connected and bounded. However, the logical (or definitional) priority need not correspond to metaphysical priority.

Let A^0 be the theory that results from our semi-Aristotelian theory by removing the axiom (2) for unrestricted fusion and adding (FINFUS).

It will prove convenient to adopt an alternative but equivalent approach, which will facilitate comparison with both the more standard theories of continuity and the

[1] At this point, it might be objected that the principle of fusions is the wrong place to look to capture the Aristotelian restrictions. The idea is that a fusion is not really an operation that produces something new. The fused thing just is the sum of its parts. It does not matter how many parts there are; since we are not producing anything with the fusion.

Although we feel no need to take sides on this metaphysical principle, we tend to agree that the Aristotelian perspective is better focused on the "potential" nature of the more standard geometric principles, such as bisection and biextension. This is done in the next section. We show there that the resulting theory is equivalent to the non-modal one developed here. Thanks to Hannes Leitgeb for pressing this issue.

[2] In some of the theories developed below, we invoke pluralities of regions. These would then be so-called "superplurals". See, for example, [97] and [80]. Readers uncomfortable with this are invited to paraphrase this talk into (finite) sets, attributes, etc.

The above referee also claims that the conceptual resources invoked here are foreign, as "Aristotle does not embrace an extensional logic. He would not, for example, recognize the 'or' with which the fusion axiom is formulated". We do not need to engage these issues here, even if we could. The present study is an attempt to use contemporary resources to get as close as we can to Aristotelian conceptions (along the same lines as, say, Michael J. White's [127]).

modal approach to be developed in the next section. The alternative is to "factor" (FINFUS) into components: two principles licensing the formation of any *finite* plurality; and, then, a principle stating that *any* plurality has a fusion—under a background assumption that all pluralities are finite. The first principle states the existence of all "singleton pluralities" and the second states that any object may be added to any plurality to yield a plurality:

$$\forall x \exists xx \forall y (y \prec xx \leftrightarrow y = x) \qquad \text{(FIN-P1)}$$

$$\forall xx \forall y \exists zz \forall w (w \prec zz \leftrightarrow w \prec xx \vee w = y) \qquad \text{(FIN-P2)}$$

The fusion principle is

$$\forall xx \exists y (y = \sum xx) \qquad \text{(P-FUS)}$$

where $y = \sum xx$ is defined as follows:

$$\forall z (z \circ y \leftrightarrow \exists x (x \prec xx \,\&\, z \circ x)).$$

Of course, we do *not* want a comprehension principle, to the effect that every non-empty predicate defines a plurality, a plural version of our semi-Aristotelian axiom of unrestricted fusions. Again, our Aristotelian presupposition is that all pluralities are finite.

With help from Dedekind, and, indeed, Euclid, we can formulate a version of this presupposition. First, the notion of *equinumerosity*, between pluralities, can either be formalized in the standard second-order way, or else, following [53], by adopting a new primitive symbol '$\approx$' subject to axioms stating that (a) $xx \approx yy$ for any one-membered pluralities xx and yy, and (b) if $aa \approx bb$ and aa' and bb' are the result of adding one object to aa and bb, respectively, then $aa' \approx bb'$. Then we could add an axiom stating that if the xx are among the yy but not all of the yy, then the xx are not equinumerous with the yy:[3]

$$\forall xx \forall yy [(\forall x (x \prec xx \rightarrow x \prec yy) \,\&\, \exists y (y \prec yy \,\&\, y \not\prec xx)) \rightarrow xx \not\approx yy]. \qquad \textit{(Part–Whole)}$$

Let A^{0^*} be the theory that results from our semi-Aristotelian theory by removing the axiom (2) for unrestricted fusion and adding FIN-P1, FIN-P2, and P-FUS as our principles for fusions, along with the other stated definitions and axioms.

Unfortunately, A^0 and A^{0^*} are far too weak. Presumably, an Aristotelian would want to assert the existence of the meet of overlapping intervals, as well as the existence of bisections, biextensions and appropriate differences. None of these can be proved in A^0 or A^{0^*}. Indeed, let X be the set of all open intervals of real numbers in the form $(a, b + \pi)$, where a and b are rational numbers. Define the fusion of a subset x of

[3] In effect, we adopt a variation on Euclid's fifth Common Notion "The whole is greater than the part". Notice that we have not *ruled out* infinite pluralities, on logical grounds alone. That, of course, cannot be done with first-order resources alone. It is consistent with the foregoing treatment to interpret the '$\approx$' relation as "finite and equinumerous".

X to be the union of x, and let Y be the set of all fusions of finite subsets of X. Say that a region a is "Left" of a region b if every member of a is less than every member of b, and let "congruent" have its usual definition, on intervals. It is straightforward to check that Y is a model of both A^0 and A^{0^*}—it satisfies our axioms governing finite fusions and all of the axioms of the above semi-Aristotelian theory except, of course, Axiom 2, the principle of unrestricted fusions. Note, for example, that the fusion of any two overlapping members of X is itself a member of X. Moreover, suppose that i and j are members of X, and that i' is congruent to i and either left-end-equivalent to j or right-end-equivalent to j. Then i' is also a member of X. So the translation of any interval in Y is also an interval in Y.

However, Y does not satisfy differences, bisections or biextensions. For example, the intervals $(0, \pi)$ and $(0, 1+\pi)$ are both in Y, but neither $(0, \pi/2)$, nor $(\pi, 2\pi)$, nor $(\pi, 1+\pi)$ are. So $(0, \pi)$ has neither a bisection nor a right-biextension, and the two intervals do not have a difference. Notice, incidentally, that if i and j are any intervals in Y, then i is *not* adjacent to j. That is, no two intervals in Y are adjacent to each other. The following is perhaps interesting:

Lemma: If follows from A^0 and from A^{0^*} that if there is at least one pair of adjacent intervals, then *every* interval has a left- and right-biextension (as in Theorem 4 of the above semi-Aristotelian theory).

Proof. Suppose that i and j are adjacent intervals. Then $i + j$ is an interval. Say that j is Right of i. Let m be any interval. Let i' be right-end-equivalent to m and congruent to i. Let k be left-end-equivalent to i' and congruent to $i + j$. Now let j' be right-end-equivalent to k and congruent to j.

Now recall that k is left-end equivalent to i', and k is congruent to $i + j$. And recall that j is adjacent to i and on its Right. And j' is right-end-equivalent to k. So k just is $i' + j'$. And so j' is adjacent to i' and on its Right. But i' is right-end-equivalent to m. So j' is adjacent to m on its right. So the right-biextension of m is an interval that is left-end-equivalent to j' and congruent to m. The left-biextension of m is obtained similarly. ■

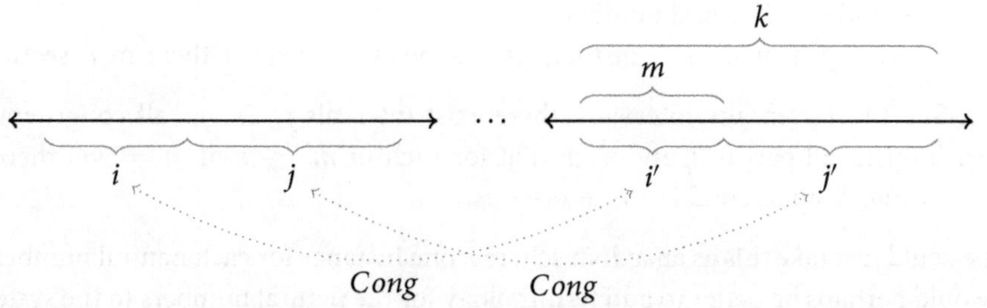

The above set Y does have meets for any pair of overlapping regions. To show that A^0 and A^{0^*} do not guarantee the existence of meets, let Z_0 be the set of all intervals (a, b), where a and b are rational numbers. Let Z_1 be Z_0 together with two more regions, $(0, 2) \cup (3, 5) \cup (6, 8) \cup (9, 11) \cup \ldots$, and $(1, 3) \cup (4, 6) \cup (7, 9) \cup (10, 12) \cup \ldots$ Finally, let Z_2 be the set of fusions of all finite subsets of Z_1, where, as in the topological models of the semi-Aristotelian system, the fusion of a set is the interior of the closure of the union of the set. Notice that Z_2 satisfies the axioms of A^0 and A^{0^*}. The two added regions overlap but they have no meet.

Let A^1 be the result of adding to A^0 an axiom of *differences*:

If $\exists z(z \circ x \,\&\, \neg(z \circ y))$ then $\exists w \forall z(z \leq w \leftrightarrow (z \leq x \,\&\, \neg(x \circ y)))$.

And let A^{1^*} be the result of adding to A^{0^*} this same axiom. As above, if $\exists z(z \circ x \,\&\, \neg (z \circ y))$ then let $x - y$ be the resulting difference.

Theorem A1: A^1 and A^{1^*} each imply the existence of biextensions of intervals and the meet of any overlapping regions.

Proof: Suppose that $x \circ y$. If $x = y$ then, of course $x \wedge y$ is just x. If $x \neq y$ then either $x - y$ or $y - x$ (or both) exist. Let z be $(x - y) + (y - x)$ (or whichever one of those that exists, if only one does). Then $x \wedge y$ is $(x + y) - z$. To obtain biextensions, let i be any interval and let j be any interval that is a proper part of i. Let j' be an interval that is left-end equivalent with i and congruent to j. Then j' is adjacent to the interval $i - j'$. The existence of biextensions thus follows from the above Lemma. ■

We still do not have bisections. To see this, let X be the set of all open intervals (a, b) where a and b are both rational numbers of the form $c/3^d$ where c is an integer and d is a natural number. Clearly, the translation of any member of X is itself a member of X. Let Y be the set of all finite fusions of members of X (where the fusion of a set is the interior of the closure of its union). It is straightforward that Y satisfies A^1 and A^{1^*}. But the interval $(0, 1)$ has no bisect, since .5 is not of the form $c/3^d$.

It would be straightforward to add an axiom of bisections:

For any interval i, there exist intervals $j < i$ and $k < i$ such that j and k are congruent, discrete, and $j + k = i$.

However, we still would not have trisections. To see this, let X' be the set of all open intervals (a, b) where a and b are both rational numbers of the form $c/2^d$ where c is an integer and d is a natural number.

So we add a principle asserting that, for each positive integer n, there are n-sections:

(n-SECT) Given any interval i, there exist intervals $j_1, \ldots, j_n$, all congruent to each other, all part of i, and such that for each $m, m' \leq n$, if $m \neq m'$, then j_m is discrete from $j_{m'}$, and $i = j_1 + \cdots + j_n$.

One could just take this as an axiom scheme, one instance for each natural number n. It would perhaps be better to add terminology for the natural numbers to the system, together with a "number of" operator. In the Aristotelian spirit, the natural numbers would be thought of as a potential infinity. Let A^2 be the resulting theory, A^1 plus (n-SECT).

We can avoid explicit talk about natural numbers by exploiting an idea that is anyway congenial to Aristotelians, namely that every positive natural number can be "instantiated" by a (finite) plurality of objects. The above axioms FIN-P1 and FIN-P2, together with some of the existence axioms above, ensure that every positive natural number is instantiated by some plurality of regions. That is, for each natural number n, there are some objects such that there are exactly n of them. Moreover, according to Aristotle *any* plurality is a finite plurality. So our axiom can be expressed as follows:

(n-SECT*) Given any interval i and any objects pp, there exist some intervals jj such that
(i) jj are equinumerous with pp,
(ii) every one of the jj is a part of i,
(iii) any two distinct members of jj are discrete,
(iv) any two members of jj are congruent to each other, and
(v) $i = \sum jj$

Let A^{2^*} be A^{1^*} plus (n-SECT*).

An Archimedean property still does not follow from A^2, nor from A^{2^*}. To see this, just let R be any non-Archimedean ordered field, and let Y be the set of fusions of finite intervals from R (where the fusion of a set is the interior of the closure of its union).

There is more than one Archimedean property that can be formulated here. Once again, Aristotle's own definition is this: "for by continual addition to a finite magnitude, I must arrive at a magnitude that exceeds any assigned limit" (*Physics*, 8, 10, 266b3). In Euclid's *Elements* (Book V, Definition 4), the Archimedean principle appears as a definition (or as a consequence of a definition): "Magnitudes are said to have a ratio to one another which are capable, when multiplied, of exceeding one another".

One Archimedean principle is a restriction of the Aristotle/Euclid principle to intervals. If we assume the natural numbers, it is straightforward to formulate this in our system. If i is an interval and n any natural number greater than zero, define the n-fold biextension of i, written ni, by recursion, as follows:

$1i = i$
$(n + 1)i$ is the fusion of ni with an interval congruent with i and adjacent on its left and an interval congruent with i adjacent on its right.

Our principle of biextensions guarantees that these intervals exist. Our first Archimedean principle is this:

(ARCH1) Let i and j be intervals. Then there is a natural number n such that j is congruent to a part of ni.

We can express a version of this Archimedean principle in the plural language, eschewing any reference to natural numbers. We define

if xx is a one-member plurality, then $xx \cdot i = i$
for any xx and any yy, if yy is the result of adding a single object to xx, then $yy \cdot i$ is the fusion of $xx \cdot i$ with an interval congruent with i and adjacent on its left and an interval congruent with i adjacent on its right.

Following our assumption that all pluralities are finite, our axioms guarantee that all of these intervals exist. The relevant Archimedean principle is this:[4]

(ARCH1*) Let i and j be intervals. Then there are some things xx such that j is congruent to a part of $xx \cdot i$.

It follows from Theorem 5 above that the semi-Aristotelian theory proves (ARCH1), and it proves a version of (ARCH1*) where the plural variables are restricted to finite pluralities.

In the present context, (ARCH1) and (ARCH1*) each rules out *intervals* that are infinitesimal, or infinite, relative to each other, but neither axiom rules out *regions* that are infinitely long, relative to intervals. Indeed, any model of the semi-Aristotelian theory is also a model of A^2 plus (ARCH1) and a model of A^{2^*} plus (ARCH1*).

Moreover, neither theory rules out the existence of intervals that are "infinitely far apart" from each other, relative to a given interval: let R and R' be two copies of the real numbers, and let us say that every member of R' is greater than every member of R. Let X be the set of all intervals (m, n) such that m and n are both members of R, or m and n are both members of R'. Let Y be the set of all finite fusions of members of X (where, again, the fusion of a set is the interior of the closure of the union). So the regions in X are finite sets of intervals. It is straightforward to verify that Y satisfies the axioms of A^2 and A^{2^*} and also (ARCH1) and (ARCH1*). Of course, Theorem 5 above rules out spaces like this, in our original "semi-Aristotelian" theory.

The following are stronger Archimedean principles, stating that if we start with an interval, and keep biextending it, we eventually include any given *region*:

(ARCH2) Let i be an interval and let k be any region. Then there is a natural number n such that $k < ni$.

(ARCH2*) Let i be an interval and let k be any region. Then there are some things xx such that $k < xx \cdot i$.

It follows that for every region m, there is an interval i such that $m \leq i$. Of course, the semi-Aristotelian theory *refutes* (ARCH2) and (ARCH2*), since that theory proves the existence of infinitely large regions, such as the entire space G. However, we think that an Aristotelian would endorse these stronger Archimedean principles.[5]

[4] To continue a theme of note 3, the conjunction of our first Archimedean principle (ARCH1), which presupposes natural numbers, with the plural axiom (n-SECT*), does rule out the existence of infinite pluralities. Suppose that there were some regions jj that are infinite in number, say of cardinality κ. Let i be any interval. It would follow, from (n-SECT*), that i has a κ-section—as strange as that may sound. That is, i is the fusion of κ-many intervals, all congruent to each other and all pairwise discrete. These κ-sects are infinitesimal relative to the given interval. It follows that there is at least one of these κ-sects such that, for each natural number n, the n-fold biextension of j is still a part of i. This violates (ARCH1).

[5] Metaphysicians have introduced the term "junk" for a sort of dual to gunk. A space is "junky" if everything in it is a proper part of something. Of course, such spaces are inconsistent with the principle of unrestricted fusions, Axiom 2 of the above semi-Aristotelian theory (at least as that is usually understood). The semi-Aristotelian space above is thus gunky but not junky (since it has a universal region G). The

Let A^3 be the theory obtained from A^2 by adding (ARCH2); and let A^{3^*} be the theory obtained from A^{2^*} by adding (ARCH2*). These, finally, are our first Aristotelian theories. It is, we think, interesting that we seem to have to add several rather different axioms in order to recover what an Aristotelian takes to be the legitimate consequences of the rejected axiom of unrestricted fusions.

One model of A^3 and A^{3^*} consists of all fusions of finite sets of open intervals in the real numbers (where the fusion of a set is the interior of the closure of its union). This is, in a sense, the intended model of each theory.

Another model consists of the regular open sets of real numbers that are bounded on both sides (i.e. are subsets of an interval). And a third consists of the regular closed sets that are bounded. To be sure, an Aristotelian would not countenance some of the members of these models. Consider, for example, the union of the open intervals $(.1, .9), (.91, .99), (.991, .999), \ldots$ That set is regular open (and its closure is regular closed), but, surely, an Aristotelian would not accept such regions.

We can rule out regions like this with an axiom stating that every region is the fusion of finitely many intervals. This is easily formulated in the plural language, recalling that we are thinking of all pluralities as finite. The optional axiom is

For any region m, there are some intervals ii such that $m = \sum ii$ (FINITE FUSION)

Another model of A^3 and of A^{3^*} (plus (FINITE-FUSION)) consists of all fusions of finite sets of open intervals whose endpoints are *rational* numbers. Yet another model consists of all fusions of finite sets of open intervals in the algebraic numbers.

So we have not ruled out countable models of our Aristotelian theories. In light of the axiomatization, this is to be expected. Every axiom of A^3 is first-order, except perhaps (n-SECT) and (ARCH2), which make direct reference to the natural numbers. So one natural background for A^3 is ω-logic, a formalism with a sort of variable ranging over natural numbers (or something equivalent), assumed to be standard. Some of the axioms of A^{3^*} are, of course, formulated in the language of plurals. A natural background framework for A^{3^*} is a plural analogue of so-called *weak second-order logic*, where the plural quantifiers are restricted to finite pluralities. The (downward) Löwenheim–Skolem theorem holds for both ω-logic and this variant of weak second-order logic (see [105], ch. 9). So we have not adopted a formalism with the resources to rule out countable models.

A fortiori, we have not ruled out models of the Aristotelian theory that are, in a sense, Dedekind incomplete (see Appendix 2.1). Indeed, we do not know how to even *state* a principle of Dedekind completeness in an Aristotelian setting. We'd have to

present, Aristotelian theories are both—a space is called "hunky" if it is both gunky and junky. See Bohn [20] and Schaffer [102]. The latter has arguments that junky spaces are metaphysically impossible. This would be interesting news to Aristotle and Euclid (and anybody else who has or had qualms about the actual existence of points and actual infinities).

be able to talk about, and have quantifiers ranging over, infinite sets of regions (or points). The same goes for stating a principle of Cauchy completeness, which requires formulating, and quantifying over, infinite sequences of regions, or intervals.[6]

In the contemporary, punctiform setting, Dedekind and Cauchy completeness capture the intuition that continua are without gaps—even point-sized gaps. Here, of course, we do not recognize points at all (at least not as "in" the space, in any sense). The closest we come to the notion of "gaplessness" is our definition of "connected", which is a property that all intervals have, by definition. Recall, from the semi-Aristotelian development, that a region x is connected if, for any regions y, z, u, if z and u are part of x and if y is between z and u, then y is also part of x. We suggest that, in the Aristotelian setting, this is all the "completeness" one can ask for.

Define a *DC-model* of A^3 and A^{3^*} (with or without (FINITE-FUSION)) to be a model in which:[7]

(i) the "regions" are sets of real numbers.
(ii) "region" m is "part" of "region" m' if and only if $m \subseteq m'$.
(iii) "region" m is "Left" of "region" m' if and only if for every $r \in m$ and $r' \in m'$, $r \leq r'$.
(iv) if a "region" m is an "interval" in the model (i.e. if it is "connected" and "bounded") then it is an interval in the real numbers (i.e. of the form (r, r'), $[r, r')$, $(r, r']$, or $[r, r']$).
(v) two such "intervals" are "congruent" just in case they have the same length. So, for example, the open interval $(3, 5)$ is "congruent" to the half-open $[7, 9)$.

Now let $\mathcal{M}$ be a DC-model of A^3 and A^{3^*} and suppose that $\mathcal{M}$ contains an "interval" whose endpoints are 0 and 1 (or, indeed, any interval whose endpoints are both rational numbers). Repeated n-sections, extensions, and translations shows that for every pair of rational numbers $r \neq r'$, $\mathcal{M}$ has an "interval" whose endpoints are r and r'. So, in a sense, the above model, in which the regions are finite fusions of open intervals whose endpoints are rational numbers, is the "smallest" DC-model.

Let m and m' be two sets of real numbers. Say that $m \cong m'$ if and only if they differ only on boundaries. More precisely, $m \cong m'$ if and only if the closure of the interior of m is identical to the closure of the interior of m'. Clearly, this is an equivalence relation (see [5], p. 236 for another, related use for this relation). Moreover, each equivalence class contains exactly one regular open set and exactly one regular closed set.

[6] Recall that, in recapturing the real numbers in our original, semi-Aristotelian framework, we take fusions of "Cauchy sequences" of intervals. Once again, this takes the sequences to be "actual", since the regions are fused into a *single* region. In the next chapter, we develop a theory with variables ranging over potentially infinite sequences. However, we think that this goes beyond Aristotelian resources, in the direction of contemporary constructive mathematics.

[7] In the remainder of this section, a word in quotes (e.g. "region", "interval", "part") refers to the interpretation of that item in the model in question. Without quotes, the word has its usual interpretation in the real numbers.

Now let $\mathcal{M}$ be any DC-model of A^3 and A^{3*}. Then $\mathcal{M}$ can contain at most one member of each equivalence class. Indeed, suppose that m and m' are both "regions" in $\mathcal{M}$ and that $m \neq m'$. Then at least one of the differences: $m - m'$ or $m' - m$ exists. So one or the other of them must contain an "interval" as a "part". So either m has an "interval" that is "discrete" from m' or else m' has an "interval" that is "discrete" from m. Either way, $m \not\cong m'$.

The Archimedean axiom entails that every region is contained in an interval. So the "largest" DC-model of A^3 and A^{3*} (without (FINITE-FUSION)) is the above interpretation in which the "regions" are regular open sets (or regular closed sets) of real numbers that are bounded on both sides. Of course, the "largest" DC-model of A^3 and A^{3*} with (FINITE-FUSION) is one where the "regions" are finite fusions of intervals.

3.4 Going Modal

The "gunkiness" Axiom 5 is that every region has an interval as a proper part. This guarantees that every model of A^3 and A^{3*} is itself infinite, in the sense that it has infinitely many regions (given the axioms of mereology). So we have not completely avoided infinity—nor do we want to. One can think of the intended domain, informally, as a "merely potential infinity". It is not clear, however, just what that amounts to. See also [81].

This section is an attempt to make more direct sense of Aristotle's claim, cited above, that "the infinite has this mode of existence: one thing is always being taken after another, and each thing that is taken is always finite, but always different" (*Physics* 3.6, 206a27–9). At least with hindsight, it is reasonably clear what an Aristotelian does *not* have: actually infinite totalities or pluralities. But what does she have instead? Just what is a merely potential infinity?

There is a closely related matter. It is generally agreed that Euclid's *Elements* captures at least the spirit of geometry during Plato's and Aristotle's period. Most of the language in the *Elements* is dynamic, talking about what a (presumably idealized) geometer can *do*. For example, the first Postulate is "To draw a straight line from any point to any point", and the second is "To produce a finite straight line continuously in a straight line". Or consider the infamous Fifth:

> That, if a straight line falling on two straight lines make the interior angles on the same side less than two right angles, the two straight lines, if produced indefinitely, meet on that side on which are the angles less than the two right angles.

As noted in Chapter 1, in Book 10, Proposition 1, an Archimedean principle, is itself cast in dynamic terms:

> Two unequal magnitudes being set out, if from the greater there be subtracted a magnitude greater than its half, and from that which is left a magnitude greater than its half, and if this

process is repeated continually, there will be left some magnitude less than the lesser magnitude set out.

Plato was critical of the geometers of his day, arguing that the dynamic language is inconsistent with the nature of the subject matter of geometry:

> [The] science [of geometry] is in direct contradiction with the language employed by its adepts . . . Their language is most ludicrous, . . . for they speak as if they were doing something and as if all their words were directed toward action . . . [They talk] of squaring and applying and adding and the like . . . whereas in fact the real object of the entire subject is . . . knowledge . . . of what eternally exists, not of anything that comes to be this or that at some time and ceases to be. (*Republic*, VII)

Of course, Aristotle rejected this Platonism, and, we think, the dynamic language actually employed in geometry better reflects his views. The matter of infinity is tied to this. For Aristotle, we never have infinite collections. What we do have are various *procedures* that can be iterated indefinitely, and we speak about what those procedures *could* produce, or what they will eventually produce. The background is dynamic and modal. In holding that the geometric procedures in question can be iterated indefinitely, Aristotle again follows the mathematical practice of the time, this time in opposition to his major opponents, the more radical atomists (see [87]), who postulate a limit to, say, bisection.

When it comes to the infinite, views like Aristotle's were standard throughout the medieval and early modern period, well into the nineteenth century. The greatest mathematical minds insisted that only the potentially infinite makes sense. Leibniz, for example, wrote:

> It could . . . well be argued that, since among any ten terms there is a last number, which is also the greatest of those numbers, it follows that among all numbers there is a last number, which is also the greatest of all numbers. But I think that such a number implies a contradiction . . . When it is said that there are infinitely many terms, it is not being said that there is some specific number of them, but that there are more than any specific number.
> (Letter to Bernoulli, [72], III 566, translated in [76], 76–7, 87)

> . . . we conclude . . . that there is no infinite multitude, from which it will follow that there is not an infinity of things, either. Or [rather] it must be said that an infinity of things is not one whole, or that there is no aggregate of them. ([73], 6.3, 503, translated in [76], 86)

> Yet M. Descartes and his followers, in making the world out to be indefinite so that we cannot conceive of any end to it, have said that matter has no limits. They have some reason for replacing the term "infinite" by "indefinite", for there is never an infinite whole in the world, though there are always wholes greater than others ad infinitum. As I have shown elsewhere, the universe cannot be considered to be a whole. ([74], 151)

For Leibniz, as for Aristotle, as for a host of others, the infinite just *is* the limitlessness of certain processes; no actual infinities exist. The only intelligible notion of infinity is that of potential infinity—the transcendence of any (finite) limit.

For at least the cases of interest here—regions, natural numbers, and the like—Georg Cantor argued for the exact opposite of this, claiming that unless it is somehow backed by an actual infinity, the potentially infinite is dubious:[8]

> I cannot ascribe any being to the indefinite, the variable, the improper infinite in whatever form they appear, because they are nothing but either relational concepts or merely subjective representations or intuitions (imaginationes), but never adequate ideas. ([27], 205, n. 3)
>
> ... every potential infinite, if it is to be applicable in a rigorous mathematical way, presupposes an actual infinite. ([28], 410–11).

We think it safe to say that this Cantorian orientation is now dominant in the relevant intellectual communities, especially concerning the mathematical domains mentioned above, with various constructivists as notable exceptions.

3.4.1 *The modality*

The task at hand is to formalize the dynamic orientation, and the concomitant notion of potentiality, by using a modal language. In contemporary jargon, we will formulate a principle that each "possible world" has access to another one, which contains all the regions of the first world and possibly some more, such as some n-sections, biextensions, and translations of the regions in the first world. But we can assume that each such "world" is finite. The total space of "worlds" is, of course, infinite, but we understand talk of possible worlds as only a manner-of-speaking, a heuristic to motivate the modal principles. The theory itself is formulated in the modal language, with the modal operators primitive.

Our "potentialist" account here takes its inspiration from Linnebo [79], which develops a modal explication of the Cantorian notion that the universe of set theory is itself potential (see note 8). Compared to that, the present situation is considerably

[8] On the surface, at least, Cantor was not consistent in his rejection of the potential infinite. Sometimes he ascribed to so-called "absolutely infinite" collections (like the ordinals) features closely analogous to those of the potential infinite. In a much quoted letter to Dedekind, in 1899, he wrote:

> [I]t is necessary ... to distinguish two kinds of multiplicities (by this I always mean definite multiplicities). For a multiplicity can be such that the assumption that all of its elements 'are together' leads to a contradiction, so that it is impossible to conceive of the multiplicity as a unity, as 'one finished thing'. Such multiplicities I call *absolutely infinite* or *inconsistent multiplicities*... If on the other hand the totality of the elements of a multiplicity can be thought of without contradiction as 'being together', so that they can be gathered together into 'one thing', I call it a *consistent multiplicity* or a 'set'. (Ewald [41], 931–2)

An 1897 letter to Hilbert is even more suggestive:

> I say of a set that it can be thought of as finished ... if it is possible without contradiction (as can be done with finite sets) to think of all its elements as existing together ... or (in other words) if it is possible to imagine the set as actually existing with the totality of its elements. (Ewald [41], 927)

Jané [66] is an insightful discussion of these themes in Cantor's thought at various stages of his career. Of course, present concern is only with collections that are, at most, the size of the continuum. Cantor considered all of those to be actual infinities (or, to be exact, he considered them to be transfinite).

simpler, since every world here is (or can be taken to be) finite. We have *no* actual infinities, or, in Cantorian terms, nothing is transfinite. Also, here we do not have to worry about higher-order resources, at least on the basic Aristotelian theory.

We must first say something about the modality that we invoke, which motivates a specific modal logic.[9] Then we show how the desired Aristotelian mathematical reasoning can be carried out in that framework. We begin with the former task. As indicated already, when explaining the modality, it will often be useful to indulge in talk about possible worlds. But, again, this is for heuristic purposes only. The modal logic that our heuristic reasoning motivates will be rock bottom, not explained or defined in terms of anything else.

A notable feature of the modal reasoning we will employ is that the domains of the possible worlds grow along the accessibility relation. So we assume:

$$w_1 \leq w_2 \rightarrow D(w_1) \subseteq D(w_2),$$

where '$w_1 \leq w_2$' says that w_2 is accessible from w_1, and for each world w, $D(w)$ is the domain of w. For present purposes, we can think of a possible world as determined completely by the regions it contains. So we can add that if $D(w_1) = D(w_2)$, then $w_1 = w_2$; and we can strengthen the above to a biconditional. However, the above conditional is sufficient for our technical purposes. We will talk neutrally about the extra regions existing at a world w_2, but not at another world w_1 which accesses w_2, as having been "generated".

As is well known, the above conditional entails that the converse Barcan formula is valid. That is,

$$\exists x \Diamond \phi(x) \rightarrow \Diamond \exists x \phi(x) \qquad \text{(CBF)}$$

This alone makes it doubtful that the modality in question can be "ordinary" metaphysical modality—whatever exactly that is. For it is widely held that there are objects whose existence is metaphysically contingent (Williamson [132] to the contrary). For example, let $\phi(x)$ say that x does not exist. Presumably, there is someone, such as Aristotle, or a given region, that might not have existed. So we have $\exists x \Diamond \phi(x)$. But then it would follow via (CBF) that it is possible for there to exist something that doesn't exist, which is absurd.

One plausible option is that the modality employed in the explication of potential infinity can be defined in terms of "ordinary" metaphysical modality. In terms of possible worlds, the relevant modality is the one that results from restricting the accessibility relation R associated with metaphysical modality by imposing the additional

[9] We make no attempt here (or anywhere else) to be faithful to everything, or even most things, that Aristotle says about modality. Michael J. White reminded us that an analysis or explanation "of the various modal principles enunciated by Aristotle (which may not form a consistent whole and indeed may involve more than one conception of the alethic modalities) is . . . notoriously difficult and vexed". We invoke some technical notions from contemporary modal logic and metaphysics in an attempt to explicate the broadly Aristotelian notion of potential infinity and, indeed, the broadly Aristotelian notion of potential existence.

requirement that domains only ever increase along the accessibility relation. More explicitly, we define:[10]

$$w_1 \leq w_2 \leftrightarrow w_1 R w_2 \,\&\, D(w_1) \subseteq D(w_2).$$

An alternative response is to sever any link between metaphysical modality and the modality invoked in explicating potential infinity. Instead, we might regard the latter as an altogether distinct kind of modality, say the logico-mathematical modality of [96] or [56], or the interpretational modality of [44] or [79]. Since our project here is to explicate a broadly Aristotelian view, it is perhaps natural to adopt the previous response, forging a connection with metaphysical modality. Nothing hangs on this, however. The formalism can characterize the modality implicitly.

3.4.2 *Brief interlude on Aristotle on mathematics (and modality)*

It is, of course (much too) much of an anachronism to wonder if Aristotle himself would accept the above principle that possible worlds grow (or do not shrink) along the accessibility relation (and, of course, this talk is only a "manner-of-speaking" for us). As noted in Chapter 1, Aristotle does say that when two continuous substances are divided, a boundary comes into existence (i.e. is made actual) and that when two such things are "put together, one boundary does not exist, but has ceased to exist" (*Metaphysics* 3.5, 1002a28–b4; see *Physics* 5, 227a10 for the definition of continuity; see [116], 11). Since, for Aristotle, points are the boundaries of line segments, when two segments are brought together, the common point ceases to (actually) exist, seemingly in contradiction with the above principle about possible worlds (if you forgive the anachronism). Of course, the present account eschews points altogether, at least as parts of anything, but that does not eliminate the issue.

It is not clear what Aristotle makes of the mathematicians' talk of geometric construction. As noted, for Aristotle, points are just the boundaries of line segments. But line segments are just the edges of plane figures, and plane figures are just the surfaces of ordinary, physical objects. Aristotle rejects the existence of a separate, Platonic realm that contains these mathematical entities. But this does not make geometry a branch of, say, mechanics. For Aristotle, the mathematician speaks of regions in a special way, not just as the relevant parts of physical objects.

The following passage from *Metaphysics* M (1077b–1078a) is more or less typical:

> [I]t is possible for there to be statements and proofs about perceptible magnitudes, but not as perceptible but as being of a certain kind. In the case of moving things there will be statements and branches of knowledge about them, not as moving but merely as bodies, and again merely as planes and merely as lengths, . . . [I]t is also true to say without qualification that mathematical

[10] This restriction can also be captured proof-theoretically. Let $\blacksquare\phi$ indicate that ϕ is metaphysically necessary. Then the relevant notion $\Box$ of necessity can be defined by letting $\Box\phi$ abbreviate $\exists xx(\forall x(x \prec xx) \land \blacksquare(Exx \rightarrow \phi))$, that is, the metaphysical necessity of ϕ, conditional on everything that in fact exists still existing.

objects exist and are as they are said to be... [T]he mathematical branches of knowledge will not be about perceptible objects just because their objects happen to be perceptible... but neither will they be about other separate objects over and above these... So if one posits objects separated from what is incidental to them and studies them as such, one will not because of this speak falsely... A man is one and indivisible as a man, and the arithmetician posits him as one indivisible and studies what is incidental to man as indivisible; the geometer on the other hand studies him neither as a man nor as indivisible, but as a solid object... That is why the geometers speak correctly: they talk about existing things and they really do exist...

See the Introduction to [2] (or [107], ch. 3, for a brief overview). The present talk of "regions" is meant in a similar, idealized spirit to Aristotle's, but we make no claims to historical accuracy.

Euclid's *Elements* does contain talk of moving figures around (although this does not occur very often, and there is no postulate that sanctions it). One prominent place is Proposition 3 of Book I. Suppose that *ABC* and *DEF* are two triangles, where the line *AB* is congruent ("equal") to *DE*; *AC* is congruent to *DE*, and the included angle *ABC* is congruent to the included angle *DEF*. The conclusion of the Proposition is that the other side *BC* is congruent to *EF*, and the triangles are themselves congruent. In the proof, Euclid begins, "for if the triangle *ABC* be applied to the triangle *DEF*, and if the point *A* be placed on the point *D* and the straight line *AB* on *DE*, ..."

There are two ways to interpret this. One is that the original triangle *ABC* is *itself* placed on top of *DEF*. So before the operation, there were two triangles, after it there is just one (as it happens). This, of course, violates the non-decreasing principle that we are adopting. One triangle simply disappears as a result of the operation.

This, however, does not fit with how the geometer treats the figures. We do not say that the original triangle *ABD* is *identical* to *DEF*, only that it is congruent to it. And, as above, Aristotle is keen to make sense of what the geometer says.

A second way to interpret the language in I 3 is that Euclid is envisioning a triangle that is congruent with the given one *ABC*, but located at the place indicated. This, in effect, invokes a two-dimensional version of the above Axiom 6, of Translation ("Given any two intervals, *i* and *j*, each is congruent both to a unique left-end-equivalent and to a unique right-end-equivalent of the other"). And it is consistent with the above principle concerning the domains of possible worlds.

We can leave it to scholars to determine which reading is right. In our reconstruction, we go for the second, and maintain the above principle that domains do not shrink along the accessibility relation.

3.4.3 *Modal logic*

Returning to the formal development, our next pressing issue concerns the right *logic* for the modality we use to explicate potential infinity. Again, it will be useful to indulge in talk about possible worlds, writing the associated accessibility relation as $\leq$. Recall that $w \leq w'$ means that we can get from w to w' by generating more regions. This motivates the following principle:

Partial ordering: The accessibility relation $\leq$ is a partial order. That is, it is reflexive, transitive, and antisymmetric.

We can also require the accessibility relation to be well-founded, especially since we are thinking of each world as finite. But nothing of substance turns on this.

At any given stage in the process of constructing regions, we will generally have a choice of which regions to generate. For example, given two intervals that don't yet have bisections, we can choose to bisect one or the other of them, or perhaps to bisect both simultaneously. Assume we are at a world w_0 where we can choose to generate regions so as to arrive at either w_1 or w_2. It makes sense to require that the license to generate a region is never revoked as our domain expands. The option to, for example, bisect a given line segment can always be exercised at a later stage.

This corresponds to the requirement that the two worlds w_1 and w_2 can be extended to a common world w_3. This property of a partial order is called *convergence* and is formalized as follows:

$$\forall w_1 \forall w_2 \exists w_3 (w_1 \leq w_3 \;\&\; w_2 \leq w_3)$$

We therefore adopt the following principle.

Directedness: The accessibility relation $\leq$ is convergent.

This principle ensures that, whenever we have a choice of regions to generate, the order in which we choose to proceed is irrelevant. Whichever region(s) we choose to generate first, the other(s) can always be generated later. Unless $\leq$ was convergent, our choice whether to extend the ontology of w_0 to that of w_1 or that of w_2 might have an enduring effect.[11]

The mentioned properties of the accessibility relation $\leq$ allow us to identify a modal logic appropriate for studying the generation of regions. Since $\leq$ is reflexive and transitive, the modal logic S4 will be sound with respect our intended system of possible worlds. As is well known, the convergence of $\leq$ ensures the soundness of the following principle as well:

$$\Diamond\Box p \rightarrow \Box\Diamond p.$$

The modal propositional logic that results from adding this principle to a complete axiomatization of S4 is known as S4.2.

As already discussed, we also have the Converse Barcan Formula, which means that S4.2 can be combined with an ordinary theory of quantification with no need for any complications such as a free logic or an existence predicate.

[11] To be sure, this principle does not hold for other sorts of modality. Suppose, for example, that I can either bake bread or bake a cake, since I have enough flour, eggs, water, etc. to do either. So there is one accessible world in which I bake bread and another in which I bake a cake. But if I choose to bake bread, it may no longer be the case that I can still bake a cake, as I may no longer have the ingredients. See Barker [11]. This does not present a problem for the present approach. In addition to the above restriction to non-decreasing domains, our first-order variables range over regions, where it does not make sense to speak of using up resources.

3.4.4 *Bridging actualist and potentialist theories*

We now turn to the second task identified above, namely the mathematical one. As observed, the existence claims of our Aristotelian theories A^3 and A^{3^*} are to be replaced by corresponding claims about possible existence—about what can be constructed or generated. How will reasoning from these modal assumptions compare with the non-modal reasoning of the previous section? Fortunately, this question admits of a systematic answer, with no need for *ad hoc* maneuvering: there are bridging principles that relate "potentialist" claims with the corresponding "actualist" ones.

Two different kinds of generalization are available in the modal framework. First, there are the generalizations expressed by the ordinary quantifiers $\forall$ and $\exists$. Since the variables range just over the ontology of the relevant world (so to speak), this is an *intraworld* form of generality. That is, a sentence in the form $\forall x\phi$, for example, is true at a world w just in case ϕ holds of all objects in $D(w)$, the domain of w. But there is also another, *transworld* form of generality available, expressed by the complex strings $\Box\forall$ and $\Diamond\exists$. These strings have the effect of generalizing not just over all entities at the relevant world, but over all entities at all worlds, or at least all accessible worlds. This idea will receive a precise statement in a "Mirroring Theorem" that we will state shortly. Loosely speaking, this theorem says that, under some plausible assumptions, the strings $\Box\forall$ and $\Diamond\exists$ behave logically just like quantifiers ranging over all entities at all (future) worlds. We refer to these strings as *modalized quantifiers*, although they are strictly speaking composites of a modal operator and a quantifier proper.

Some definitions: if ϕ is a non-modal formula, its *potentialist translation* $\phi^\Diamond$ is the formula that results from replacing each ordinary quantifier in ϕ with the corresponding modalized quantifier. That is, '$\forall x$' is replaced by '$\Box\forall x$'; and '$\exists x$' is replaced by '$\Diamond\exists x$'. Say that a formula, in the modal language, is *fully modalized* just in case it is in the form $\phi^\Diamond$, where ϕ is a formula in the non-modal language.

Say that a formula ϕ, in the modal language, is *stable* if the necessitations of the universal closures of the following two conditionals hold:

$$\phi \rightarrow \Box\phi$$
$$\neg\phi \rightarrow \Box\neg\phi$$

Intuitively, a formula is stable just in case it never "changes its mind", in the sense that, if the formula is true (or false) of certain objects at some world, it will remain true (or false) of these objects at all "later" worlds as well.

Let $\vdash$ be the relation of classical deducibility[12] in a language L. Then let $\vdash^\Diamond$ be deducibility in the modal language corresponding to L, by $\vdash$, S4.2, and the stability

[12] In general, we mean classical deducibility without the use of any plural or higher-order comprehension axioms. Fortunately, this issue does not arise here, since our theories, A^3 and A^{3^*}, do not have comprehension principles. The natural background for A^{3^*} is a plural language, but we only have principles that guarantee finite pluralities, and we have no comprehension axiom or scheme for those.

axioms for all atomic predicates of L. We are now ready to state the mentioned theorem.

Mirroring Theorem: For any formulas $\phi_1, \ldots, \phi_n, \psi$ of L, we have:

$$\phi_1, \ldots, \phi_n \vdash \psi \quad \text{iff} \quad \phi_1^{\Diamond}, \ldots, \phi_n^{\Diamond} \vdash^{\Diamond} \psi^{\Diamond}.$$

See [79] for a proof.[13]

The Mirroring Theorem tells us that, if we are interested in logical relations between fully modalized formulas in a modal theory that includes S4.2 and the stability axioms, we may delete all the modal operators and proceed using the ordinary non-modal logic underlying $\vdash$. Thus, under the assumptions in question, the composite expressions $\Box\forall$ and $\Diamond\exists$ behave logically just like ordinary quantifiers, except that they generalize across all (accessible) possible worlds rather than a single world. This provides the desired bridge between actualist and potentialist theories. Of course, the modal languages are at least potentially richer, as they can express formulas that are not in the range of the potentialist translation.

3.4.5 *Why the Mirroring Theorem is available*

Provided that the assumptions of the Mirroring Theorem are satisfied, this bridge makes it straightforward to "potentialize" the Aristotelian theories A^3 and A^{3^*} developed above. The theorem will ensure that any reasoning from our actualist Aristotelian theories can be duplicated in the potentialist setting. All that remains is therefore to verify that the Mirroring Theorem is in fact available.

We have already argued (section 3.4.3) that the modal logic is (at least) S4.2. The first task is to defend the stability of our primitive predicates: $=$, $\leq$, L, *Cong*, and, in A^{3^*}, $\prec$. We begin with the geometric primitives.

Recall that $L(x,y)$ means that region x is completely to the left of region y. The stability of this predicate comes to

$$L(x,y) \to \Box L(x,y)$$
$$\neg L(x,y) \to \Box\neg L(x,y)$$

These principles would obviously be false if the variables were given unrestricted range, and included non-mathematical contexts. Consider, for instance, two sticks on a line. Even if one stick is (completely) left of the other, from some perspective, it might have been the other way round (and we can make it the case that it is the other way around by moving one of the sticks). But recall that here, our (first-order)

[13] A version of the Mirroring Theorem can be proved for an intuitionistic context as well. As usual, define a formula Φ to be *decidable* in a given theory if the universal closure of $(\Phi \vee \neg\Phi)$ is deducible in that theory. The premises of the intuitionistic Mirroring Theorem are: Let $\vdash$ be the relation of intuitionistic deducibility in a language L, and let $\vdash^{\Diamond}$ be deducibility in the modal language corresponding to L, by $\vdash$, S4.2, the stability axioms for all atomic predicates of L, and the decidability of all atomic formulas of L. The conclusion is then the same as in the text. This may prove important for those who hold that only intuitionistic logic is appropriate for the Aristotelian theories. See the next chapter.

variables range over *regions*. And it is plausible to take regions to be individuated, at least in part, by their geometrical properties. Thus, when one region is (or is not) entirely left of another, there is no construction we can undertake that would change this fact. In effect, regions do not move (at least not relative to other regions) during any process of mathematical construction (as per the discussion at the end of section 3.4.2). This motivates the pair of stability principles displayed above. The case for the stability of *Cong* is entirely analogous. If two intervals are congruent (or not), then no construction can change that. Constructions do not change the length of intervals; they only produce new regions.

The stability of identity was famously defended by Kripke [69], and we can simply adopt his argument.[14] In light of the foregoing, however, the case here is especially strong, and does not depend so much on intuitions as to what is metaphysically possible concerning things like people, sticks, and houses. If two *regions* are identical (or distinct), then, intuitively, they remain identical (or distinct) even as various constructions are performed. The same goes for the stability of $\leq$, the mereological parthood relation.

We adopt the policy (or view) of Linnebo [79] (cf. also Uzquiano [122]) that pluralities are determined entirely by their "members". So if an object a is (resp. is not) one of the bb in one world, then a is (resp. is not) one of the bb in every world that contains both a and the bb. This policy is especially clear in the present, Aristotelian context, since we only envision finite pluralities. We stipulate that such pluralities are "individuated" by the objects they "contain".

That takes care of the stability of the primitives.[15] Note that in the development of the "potentialist" theory, we should formulate everything using only the primitives, eschewing the defined terms. Alternately, the various definitions could themselves be "modalized". For example, in the above semi-Aristotelian and Aristotelian theories, we defined two regions x, y to overlap, written $x \circ y$, just in case there is a region that is part of both: $\exists z(z \leq x \,\&\, z \leq y)$. We could say that here, too, but then overlap would sometimes depend on whether a common region happened to be constructed, and so overlap would not be stable. Instead, we should say that $x \circ y$ holds just in case there *could be* a region that is part of both: $\Diamond\exists z(z \leq x \,\&\, z \leq y)$. Similarly, to say that x and y are discrete, written $x|y$, is not merely to say that there do not happen to be any regions that are part of both, but to say that there *cannot* be any regions that are part of both: $\neg\Diamond\exists z(z \leq x \,\&\, z \leq y)$. That is, two regions are discrete just in case no

[14] We note, however, that Kripke's derivation of negative stability

$$x \neq y \rightarrow \Box x \neq y$$

uses the Brouwerian axiom $\phi \rightarrow \Box\Diamond\phi$, which is not available in our logic S4.2.

[15] As in note 13, in an intuitionistic setting, we'd also have to establish the decidability of the atomic predicates. In particular the mirroring depends on whether it is decidable whether two given regions are identical, whether one of them is a part of the other, whether one of them is Left of the other, and whether two intervals are congruent. Given the connection between these relations and ordinary geometric construction, we take this to be unproblematic.

matter how much construction we do, there will never be a region that is part of both of them.

It is, of course, a routine syntactic matter to "modalize" the other definitions. We give ordinary language glosses on some of them, to help motivate the present orientation. Recall that a region x is *connected* if $\forall y, z, u[z, u \leq x \,\&\, Betw(z, y, u) \rightarrow y \leq x]$. Modalized, this says that necessarily, for any regions y, z, u, if y and z are parts of x and if u is Between y and z, then u, too, is part of x. A region is *bounded* if there could be a region to its left and there could be a region to its right. Two regions are *adjacent* if there cannot be a region that is part of both and there cannot be a region that is between them. Finally, recall that two intervals j, k are *left-end-equivalent* if $\exists p[p \leq j \,\&\, p \leq k \,\&\, \nexists q(\{q \leq j \vee q \leq k\} \,\&\, L(q, p))]$. Modalized, this says that there could be a region p that is part of both, such that there cannot be a region that is part of either j or k that is also Left of p. We take it that these are all what we want.

Our final task is to verify that the potentialist translations of the axioms of A^3 and A^{3^*} are acceptable to the potentialist. The potentialist translation of axiom 1a says that necessarily, $\leq$ is a partial order. For instance, the transitivity axiom is necessitated:

$$\Box\forall x\Box\forall y\Box\forall z(x \leq y \,\&\, y \leq z \rightarrow x \leq z) \tag{1}$$

This is acceptable to the potentialist for much the same reason as the stability axioms considered above; and the same goes for the reflexivity and antisymmetry claims.

The next axiom is 1b, which says that $x \leq y$ iff anything that overlaps x also overlaps y. If we unpack the definitions, as above, and then modalize, the result is the necessitation of the universal closure of the following:

$$x \leq y \leftrightarrow \Box\forall z(\Diamond\exists w(w \leq z \,\&\, w \leq x) \rightarrow \Diamond\exists w(w \leq z \,\&\, w \leq y)).$$

This is indeed acceptable to the potentialist: x is part of y if and only if any region z that we may generate possibly has a joint part with x only if that same z possibly has a joint part also with y.

Recall that, in effect, in our semi-Aristotelian theory A^3, the unrestricted fusion axiom has been replaced by (FINFUS), which states that any two regions have a fusion:

$$\forall u\forall v\exists x\forall y[y \circ x \leftrightarrow (y \circ u) \vee (y \circ v)].$$

Modalized, this comes to:

$$\Box\forall u\Box\forall v\Diamond\exists x\Box\forall y[y \circ x \leftrightarrow (y \circ u) \vee (y \circ v)].$$

In words, this says that, necessarily, for any regions u, v, there could be a region x (i.e. $u + v$), such that, necessarily, any region can share a part with v just in case either that region can share a part with u or that region can share a part with v. This, we submit, is what we want.

The alternate system A^{3^*} has, instead, an axiom, which states that any plurality—which according to our Aristotelian will be finite—has a fusion. The potentialist translation of this is

$$\Box\forall xx\Diamond\exists y\,(y = \textstyle\sum xx)^{\Diamond} \qquad \text{(P-FUS}^{\Diamond}\text{)}$$

where $(y = \sum xx)^{\Diamond}$ in turn unpacks as:

$$\Box\forall z(\Diamond\exists w(w \leq z \,\&\, w \leq y) \leftrightarrow \Diamond\exists x(x \prec xx \,\&\, \Diamond\exists w(w \leq z \,\&\, w \leq x))).$$

Although this latter formula is quite a mouthful, we contend that, from the potentialist's point of view, this is a perfectly plausible definition of the claim that y is the fusion of xx. Accordingly, (P-FUS$^{\Diamond}$) says what the potentialist wishes to say, namely that necessarily, any (finite number of) regions possibly have a fusion.[16]

Recall our axioms (FIN-P1) and (FIN-P2) that govern finite pluralities. The first says that for each object x there is a plurality that contains just x, and the second states that the result of adding an object to a plurality results in another plurality. The modalizations of these are, of course, acceptable. We could just assert the necessitations of these. For example, we don't have to think that the existence of a plurality that contains just x requires a separate construction. So we could say, instead, that necessarily, for each region x there is a plurality that contains just x.

Consider, next, our (Part–Whole) axiom that attempts to state the finitude of each "plurality". Modalized, this axiom says that, necessarily, if the xx are among the yy, but the xx are not all of the yy, then the xx are not equinumerous with the yy. Again, this seems correct (recalling that we envision only finite pluralities here).

As it happens, we can replace (FIN-P1) and (FIN-P2) with the (necessitation of the) ordinary (non-modal) plural comprehension scheme:

$$\exists y\phi(y) \rightarrow \exists xx\forall y(y \prec xx \leftrightarrow \phi(y)) \qquad \text{(P-Comp)}$$

for any ϕ without xx free. Notice, first, that, given our interpretation of the pluralities, the instance of (P-Comp) in which $\phi(y)$ just is $y = y$ entails that, in effect, the domain of each world is finite—just as desired. Second, if every "world" is indeed finite, then there is no difference between the pluralities defined by any non-empty condition $\phi(y)$ and those defined by finitely many steps of adding one object.

It is important not to confuse the non-modal plural comprehension scheme (P-Comp) displayed above with *the potentialist translation* of the same scheme:

$$\Diamond\exists y\phi(y) \rightarrow \Diamond\exists xx\Box\forall y(y \prec xx \leftrightarrow \phi^{\Diamond}(y)) \qquad \text{(P-Comp}^{\Diamond}\text{)}$$

This we do not want. The instance of this scheme where ϕ is $y = y$ is false for our Aristotelian: for there will *never be* a plurality—and thus a completed collection—of all possible self-identical objects. The principle P-Comp$^{\Diamond}$ would yield a modalized

[16] To recall the theme of note 1, it might be questioned whether the innermost 'possibly' that figures in this claim is appropriate. When some regions have been generated, have we not thereby also generated their fusion, such that no *further* generation is needed in order to bring this fusion into existence? We leave this for metaphysicians to ponder. For our purposes, it suffices to observe that either choice will sustain the technical argument in which we are engaged. We therefore proceed as in the main text, as this is weaker and thus less controversial, and also technically more elegant.

version of the unrestricted fusion principle, Axiom 2, of the semi-Aristotelian theory. Of course we reject this—it makes our potential infinities actual, and our Aristotelian assumes that *no* potential infinities can be completed. P-Comp$^\diamond$ cannot be obtained from P-Comp, as can be seen with a Kripke structure in which every "world" is finite.

We conclude this development with glosses on the modalized versions of the remaining axioms. First, the axioms to the semi-Aristotelian theory that carry over:

Axiom 3: Necessarily, if a region x is either Left or Right of a region y then x and y are discrete (there cannot be a part of both). And, necessarily, x is Left of y just in case, necessarily, every part of x is Left of every part of y.
Axiom 4 (Dichotomy): Necessarily, if i and j are discrete intervals, then either i is Left of j or i is Right of j.
Axiom 5 (Gunkiness): Necessarily, any region can have an interval as a proper part.
Axiom 6 (Translation): Necessarily, for any intervals i, j, there could be an interval k that is congruent with i and right-end-equivalent with j; and there could be an interval k' that is congruent with i and left-end-equivalent to j.
Axiom 7 (Additivity): Necessarily, for any intervals i, j, i', j', if i is Adjacent to j and i' is Adjacent to j', and if i is congruent to i' and j is congruent to j', then, necessarily, the fusion of i and j is congruent to the fusion of i' and j'.

This last is, again, a mouthful, but we take it that it is clearly acceptable.

Here are the glosses on the remaining axioms of A^3 and A^{3^*}:

Differences: Necessarily, for any regions x, y, if there could be a region that overlaps x and is discrete from y, then there can be a region w such that, necessarily, any region is a part of w just in case it is part of x and discrete from y.
(n-SECT): Necessarily, for any interval i there can be intervals $j_1, \ldots, j_n$ which are congruent to each other and pairwise discrete, such that i is the fusion of them.
(n-SECT*): Necessarily, for any interval i and any objects pp, there could be some intervals jj such that the jj are equinumerous with the pp, the jj are pairwise discrete, the jj are all congruent to each other, and i is the fusion of the jj.
(ARCH2): Necessarily, for any interval i and any region k, there is (or could be) a natural number n such that k is part of an n-fold biextension of i.
(ARCH 2*): Necessarily, for any interval i and any region k, there could be some things xx such that k is part of an xx-fold biextension of i.

Recall, finally, that we introduced, as a possible new axiom, a statement (FINITE FUSION) that every region is the fusion of finitely many intervals:

For any region m there are some intervals ii such that $m = \sum ii$.

Modalized, this says that, necessarily, for any region m there could be some intervals ii such that m is the fusion of the ii. This seems plausible, or at least as plausible as (FINITE FUSION). In fact, one might even strengthen this to a statement that necessarily, for any region m, there *are* some intervals ii such that m is the fusion of the ii. In terms of possible worlds, this last is a requirement that if a region m exists at a world w, then w also contains finitely many intervals whose fusion is m. This is of a piece with the thesis, broached in section 3.3, that intervals are more basic than general regions.

3.5 Looking Ahead

We began this chapter with a brief overview of the "semi-Aristotelian" theory of a one-dimensional continuum developed in the previous chapter. That theory made heavy use of "actual" infinity, through an axiom that any regions have a mereological fusion. We then developed a couple of more Aristotelian theories, that make no use of infinite fusions. We also presented a modalized version of that theory, which, we submit, captures the "potential" nature of Aristotelian (and Euclidean) construction, and we applied a version of Linnebo's [79] "Mirroring Theorem", to show that our two theories are essentially equivalent, with the modal one more expressive.

Our next task is to introduce "points" or analogues of real numbers, into the Aristotelian framework, as a sort of superstructure, along similar lines to how that is done in the previous chapter, a kind of Whitehead-style "extensive abstraction". The plan is to present a theory of potentially infinite sequences of regions. In line with the Aristotelian, we will not identify or associate each such sequence with a region. Instead, we theorize about the sequences directly. This, however, goes beyond Aristotelian resources, in the direction of modern constructive mathematics. It turns out that, unlike the foregoing semi-Aristotelian accounts, the pointy superstructure we obtain is very different from the underlying, gunky space. This will be the business of the next chapter.

We turn now to a theory that, in an intuitive sense, lies between the semi-Aristotelian and the more Aristotelian theories.

3.6 Predicative, Regions-based Continuum

In a sense, predicative mathematics sits in between the semi-Aristotelian perspective of the previous chapter and that of the "more Aristotelian" systems developed just above. Like the Aristotelian, the predicativist balks at the unrestricted use of actual infinity in the semi-Aristotelian framework, but she does not reject the actually infinite outright. There is no objection to actual infinity as such, provided that it is countable and specifiable in a predicatively acceptable way.

It might be added that the predicativist has no particular objection to punctiform systems of geometry or topology either. As predicative mathematics is usually formulated, one begins with first-order logic and axioms for the natural numbers, along now standard lines. Continua are then constructed or, perhaps better, defined, in stages: one introduces the (signed) integers, then the rationals (e.g. via pairing of integers), and then the reals via Cauchy sequences of rationals or Dedekind cuts therein. Unlike the contemporary, Dedekind–Cantor perspective, however, the predicativist requires explicit formulae, in a certain syntactic form, for describing Cauchy sequences or Dedekind cuts.

At the "first" stage beyond the rational numbers, we only consider formulas with bound variables restricted to the rational numbers. Formulated set-theoretically, such formulae provide instances of comprehension and separation axioms. At the next stage, the predicativist considers formulas with bound variables ranging over real numbers defined at the first stage. And on it goes.

Since only countably many formulae are available at any stage of the construction, predicativist mathematics works only with the countably infinite, forswearing axioms (such as that of power sets) that imply the existence of uncountable totalities. In the case of the real numbers, one may develop a hierarchy of richer and richer systems as in ramified type theory, allowing at any stage quantifiers over reals specified at earlier stages in the defining formulae. Crucially, one lacks the machinery to specify the union of all such stages. A certain countable limit ordinal, Γ_0, is usually taken to designate a limit to the available stages, but even to recognize this limit, one steps outside the predicativist framework proper (see [42], [43]).

One who is familiar with the development of classical number systems from a logicist basis, à la Frege or Russell, may wonder about the decision to start with *sui generis* axioms for the natural numbers: for on those accounts, the class of the naturals is explicitly *defined* as the minimal closure of (the singleton of) the initial number designated '0' or '1', under the (defined) operation of successor—and this is a paradigmatically impredicative construction. So isn't the predicativist program, as usually presented, on shaky ground? In response, one may begin with an elementary theory of finite sets over an axiomatically provided, minimally structured, countable domain (e.g. a pairing function and an urelement under pairing), thereby recovering a natural-numbers system (with induction for classes specified predicatively relative to finite sets), along with a proof of categoricity of the axioms and of recursion theorems, and a proof that the system is conservative relative to first-order Peano Arithmetic (again, see [42], [43]). This will prove useful in what follows here.

The goal here is to present a predicative, regions-based continuum. The situation is complicated by the fact that a theory of a regions-based continuum is, from the outset, aiming to describe an uncountable structure, one that is not predicated on, or constructed from, a prior countable one (such as the rational numbers). The semi-Aristotelian development of the previous chapter begins with first-order logic enriched with a logic of plurals (or something equivalent), along with axioms of classical mereology, including an unrestricted comprehension scheme for plurals (or equivalent), and an axiom asserting the existence of the mereological sums or "fusions" of any (all) such pluralities. Such axioms are naturally regarded as impredicative and so must be restricted somehow. Sticking again with the one-dimensional case, the task at hand is to determine what sort of predicative theory can be developed on the more restrictive basis and how that compares with the original classical theory.

3.6.1 *Predicative adjustments*

The places where impredicativity threatens the semi-Aristotelian development are essentially the same as those where the Aristotelian demurs. Recall that there were several places, in the Aristotelian development, where we took the fusion of some regions, without bothering to check whether the regions, so fused, are finite in number. The predicativist also balks at those places, since we did not check to make sure that we have a predicative definition of the relevant plurality. In other places, we explicitly took the fusion of infinitely many regions. Typically, these are also impredicative, often defined as minimal closures. As it happens, the predicativist can recapture some of those results.

As in the Aristotelian development, the troubling results, from the semi-Aristotelian development, are these:

(1) specification of arbitrary meets and differences among regions (via Axiom 2);
(2) specification of the universal region, G as the (unrestricted) fusion of *all* regions;
(3) in Theorem 4, the derivation of the existence of biextensions of any given interval (implying the bi-infinity of the space);
(4) the construction of the minimal closure of $\{i\}$ under iterated biextension (Lemma 3), a paradigmatically impredicative construction; this yielded
(5) the Archimedean property;
(6) the proof of Theorem 6, existence of bisections of intervals;
(7) the proof of Theorem 7, Dedekind completeness of the interval structure.

Recall that, in the Aristotelian theories, items (1), (3), (5), and (6) are added in, by hand, using new axioms. The Aristotelian rejects the existence of a universal region (2), and the relevant minimal closures (4) outright. Even the *statement* of Dedekind completeness (7) takes us beyond Aristotelian resources.

The task here is to formulate a regions-based theory that addresses the relevant issues as successfully as possible. The matter is complicated by the fact that, classically speaking, the individuals are regions of a geometric space, and, even well-behaved regions such as intervals, are uncountably many and so cannot enter into a predicatively acceptable totality or operation such as fusion. Thus arises the problem over the universal region G (2), and the break down of proofs of the existence of biextensions and bisections (problems (3) and (5)).

Since we are starting without any arithmetic axioms at all, it is useful to try to adapt the machinery of [42], [43] in their "predicative foundations of arithmetic", where axioms on ordered pairing provided for a countable set of individuals. Then "finite set" of such individuals is taken as primitive, subject to uncontroversial further axioms, and supplemented with a comprehension principle for "classes" of individuals (in which only individual and finite-set bound variables occur in the defining formulae).

Hence the designation 'EFSC' of [42], [43], for "elementary theory of finite sets and classes".

In the present, more geometric setting, we will combine these ideas with axioms of atomless mereology (with $\leq$, part of, as primitive) and further axioms governing the special regions called "intervals", which enable us to dispense with primitive pairing. Instead of finite sets and classes of individuals, we will use plural variables ff, gg, hh, etc., for finitely many regions or intervals, and xx, yy, zz for general plural reference; and we write $x \prec ff$, $y \prec xx$, etc. to express "x is one of, or is among, the ff", "y is one of, or is among, the xx", etc.[17]

Our axioms start off the same as for the semi-Aristotelian theory described above, but we replace the unrestricted fusions Axiom 2 initially to allow only fusions of finitely many regions—as in the Aristotelian development. Later we broaden this to include certain countable fusions—those with proper definitions.

The other axioms of the semi-Aristotelian theory, through 7, carry over intact. Recall Axiom 5, the one that renders the system "gunky":

$$\forall x \exists j[Int(j) \,\&\, j < x].$$

Since the background logic is unfree, we assume the existence of at least one region. Axiom 5 thus gives us at least one interval. We designate one such interval to serve as a unit, for the purpose of a representation of the integers and, eventually, the rationals and predicatively specifiable reals.[18]

As in the Aristotelian theory, we adopt another Axiom, 8, for differences:

$$\text{If } \exists z(z \circ x \,\&\, \neg(z \circ y)) \text{ then } \exists w \forall z(z \leq w \leftrightarrow (z \leq x \,\&\, \neg(x \circ y))).$$

Then Theorem A1 above gives the existence of meets and biextensions.

Iterated biextensions, left and right, starting with our arbitrarily assumed unit interval, generate a bi-infinite linear structure isomorphic to the integers. In what follows, we thus assume such a bi-infinite sequence of sequentially adjacent intervals, all congruent to our unit, and we will designate these *canonical intervals*.

As in the Aristotelian system, we cannot derive the existence of bisections (Theorem 6), as that involves arbitrary fusions, so instead we postulate it:

Axiom 9 *Bisections*: $\forall j \exists! k, m[Cong(k, m) \;\&\; Adj(k, m) \;\&\; j = k + m]$, i.e. j is the fusion of k and m.

Finally, we have axioms governing the plural variables, both finitary and general:

Axiom 10 *Singleton finite plurality*: $\exists ff \exists! j[j \prec ff]$.

[17] Recall that one of the Aristotelian theories above invoked finite pluralities. There was no need, there, to invoke general pluralities, since the Aristotelian does not recognize anything beyond the finite.

[18] If one prefers a free logic, then we'd need an axiom asserting the existence of an interval (or, given Axiom 5, a region).

Axiom 11 *Adjunction:* $\forall ff \forall a \exists gg \forall x[x \prec gg \leftrightarrow x \prec ff \vee x = a]$.

Axiom 12 *Finite Separation:* $\forall ff \exists gg \forall x[x \prec gg \leftrightarrow x \prec ff \wedge \Phi(x)]$, where Φ lacks free 'gg' and lacks bound general plural variables. (Thus, the individuals among the (finitely many) ff satisfying such Φ are finitely many.)

Note that we do not need an axiom of induction governing the finite plural variables (the plurals analogue of finite-set induction), because the machinery so far already suffices to define minimal closures of finitely many given objects under the operations we need, providing theorems of induction for the formulae we need (cf. [42], [43]).

What about a comprehension principle for general plurals? Here, as already noted, we have to be careful lest we commit the predicativist to uncountable totalities. Our strategy will be to introduce certain interval sequences, already furnished by the axioms presented thus far, and then to allow into our general plurals comprehension axioms formulas with bound variables ranging over these intervals. They come in two sorts, depending on the operations of biextension and bisection, respectively, leading to them. The first sort was already described following the above introduction of biextensions: we're given a starting unit interval, call it j_0; then repeated applications of biextension right and left yield a bi-infinite sequence of mutually congruent intervals of the form

$$\ldots j_{-3}, j_{-2}, j_{-1}, j_0, j_{+1}, j_{+2}, j_{+3}, \ldots$$

with each interval adjacent to the ones immediately flanking it. (We designate this doubly infinite sequence with the plural constant $jj = \langle j_i \rangle$.) More formally, our biextensions theorem (A1) furnishes us with two 1-1 functions biext_L and biext_R yielding for any argument j_i its unique left-biextension and its unique right-biextension, respectively. Then we can define the members of the above displayed canonical interval sequence by requiring that any finite plurality ff such that (1) $j_{+k} \prec ff$ and (2) if $j_i \prec ff$, then $\text{biext}_R^{-1}(j_i) \prec ff$, also satisfies that $j_0 \prec ff$, and likewise for any of the j_{-k}, *mutatis mutandis*. These are our canonical intervals.

Next we specify intervals that can be called *binary fractions* of canonical intervals. These result from iterated bisection of subintervals, left or right at each stage, beginning with any of the canonical intervals. As in the case of biextensions, Axiom 9 on bisections furnishes two 1-1 functions, bisec_L and bisec_R, yielding for each interval j, its unique left half, j_L^{-2}, and its unique right half, j_R^{-2}, respectively. Thus, any sequence $\sigma = \langle \sigma_i \rangle$ obtained by iterating these functions is determined by requiring that any finite plurality ff such that (1) $\sigma_i \prec ff$ and (2) if $k \prec ff$, then either $\text{bisec}_L^{-1}(k)$ or $\text{bisec}_R^{-1}(k) \prec ff$, also satisfies that one of the canonical intervals $jj \prec ff$.

We are now in a position to state our comprehension principle governing general plural variables and fusions pertaining to them:

Axiom 13 *Plurals Fusion Comp:* $\exists u \Phi(u) \rightarrow [\exists x \forall y \{y \circ x \leftrightarrow \exists z(\Phi(z) \,\&\, z \circ y)\}]$, where Φ lacks free x and *the bound individual or finite or general plurals variables of* Φ *are restricted to canonical intervals or binary fractions thereof.*

Applying this in the case that $\Phi(u)$ is "u is a canonical interval of the sequence jj (displayed above)", we have the existence of the fusion of the jj. To recover the classical characterization of G, we need to show that this is the same as the minimal closure of $\{j_0\}$ under iterated biextension, despite the impredicativity of the classical definition of that minimal closure as the intersection of all closures.

Here is where the method of [42], [43] can be adapted so that we can bypass the classical definition. In its stead, we simply require that the intervals starting with j_0 up to any j_i—proceeding in the sense of $+$ or $-$ of i—form a finite plurality. This suffices to establish isomorphism of the sequence j with the integers, which collectively satisfy the minimal closure condition. Thus, just as [42], [43] recover the existence and uniqueness (up to isomorphism) of the natural-numbers-structure from the requirement that all initial segments of their defined privileged sequence (obtained by iterating the pairing operation, beginning with the posited urelement) be finite, so that requirement applied in the positive and negative directions of j suffices for uniqueness up to isomorphism of our bi-infinite linear interval structure.

We are not quite done, however, for we have yet to derive the Archimedean property. The above shows that the fusion of the sequence jj is part of the fusion of all intervals that are either to the right or to the left of our initial unit interval, j_0, but we need that the latter fusion be *identical* to the fusion of the jj. But that follows by the same proof given in the classical case, applying the Translation axiom, as no impredicative definitions arise there. Thus we have the identity; G is the fusion of the canonical interval sequence jj.

Furthermore, we observe that, with the machinery of binary fractions of (lengths of) canonical intervals together with mereology and plurals, the notion of *Cauchy interval sequence* is predicatively acceptable, which opens the way to introducing—as superstructure over the interval structure of G—a hierarchy of ever richer real-number structures, as in standard predicative analysis.

What about the *completeness* of the predicative continua, either at the geometric level of intervals or at the superstructural level? This depends on the notion of "completeness" involved. As already noted, *Dedekind completeness*—the least upper bound principle applied to *arbitrary* bounded pluralities of real-length intervals—is paradigmatically impredicative, just as it is in the case of numbers. Classical theorems of analysis that use this principle must be derived by other means, or else the statements of those theorems, such as the intermediate-value theorem, must be suitably altered. However, a restricted least-upper-bound theorem for (properly definable) bounded sequences of canonical real-length intervals is easily derivable. One starts by taking the fusion of the given bounded sequence.

The situation with *Cauchy completeness*, that every Cauchy sequence of reals (or real-length intervals) converges to a real (real-length interval) of the space, is favorable both to the constructivist and the predicativist, the main difference residing in their different criteria for what counts as a "Cauchy sequence of reals (or real-length intervals)". Classically, one allows arbitrary Cauchy sequences, while the constructivist demands that a function from ε to $N(\varepsilon)$, the term of the sequence beyond which differences must be within ε, be a *constructive* function. Similarly, the predicativist requires that such a function be predicatively definable.[19] Apart from such differences in the meanings of their terms, however, the argument for Cauchy completeness goes through. Here is a simple argument applicable to any stage of a hierarchy of predicative reals (or real-length intervals, which should be understood whenever we refer simply to reals, in the following).

Theorem P1 (*Cauchy Completeness*): *Any Cauchy sequence, $\rho = \langle r_i \rangle$, of predicative reals converges.*

Proof. Each predicative real r_i is determined (up to equivalence based on co-convergence) by a Cauchy sequence, $q_i = \langle q_{ij} \rangle, j = 1, 2, 3, \ldots$ of rationals (rational-length intervals). Without loss, we consider the unit interval and assume that each r_i of the sequence ρ is presented as a row, $r_{ij}, j = 1, 2, 3, \ldots$, in decimal notation, all such rows forming a countably infinite matrix M with entries r_{ij}. It suffices to transform M into a corresponding matrix M^q with rows representing rationals, q_i, such that the sequence of those rows, $\langle q_i \rangle$ co-converges with $\langle r_i \rangle$. This is accomplished simply by setting the entries of each row equal to those of the rational sequence $\langle q_{ij} \rangle$ determining r_i, for those values of $j \leq i$, setting $q_{ij} = 0$ for all $j > i$. It is easily checked that this transforms M to M^q identical to M on the half left of the diagonal, including the diagonal entries, with 0's everywhere right of the diagonal, such that the sequence of rows $q = \langle q_i \rangle$ is co-convergent with $\rho = \langle r_i \rangle$. But by construction, $\langle q_i \rangle$ defines a real, as the limit of both the rational sequence q and the real one ρ from which it is derived. ■

Recall that, in the Aristotelian system, we had to add an axiom for n-sections of intervals (where n is a positive integer). Here we just add an axiom for bisections. The existence of n-sections follows from Theorem P1, as it is straightforward to define, for any given interval i, a predicative "Cauchy sequence" of intervals that "converges" to the left-most n-section of i.

3.6.2 *Conclusion*

The above predicative theory of a regions-based continuum goes quite far, as indicated, including the derivation of the Archimedean property from Translation, using techniques of [42], [43]. But, lacking unrestricted fusions, it has to postulate differences and bisections outright, rather than deriving them; and it fails to derive Dedekind completeness. As to the extent of the intervals recognized, the predicativist could go on to introduce a hierarchy of increasingly rich theories of countably many real-length intervals, mimicking what it already does to approximate the classical

[19] To be sure, the predicativst cannot *state* this requirement, since the notion of predicative definability is not itself predicatively definable. See [42], [43].

point-based continuum. But here the loss seems greater: in the point-based classical theory, Dedekind completeness is usually an extra axiom, one that the predicativist must forgo; but the regions-based classical theory (our "semi-Aristotelian" one) derives that principle as a theorem, whereas it remains out of reach for the predicativist. In sum, unrestricted, impredicative fusions, like its close cousin, unrestricted second-order comprehension, has great unifying power that predicative mathematics cannot access.

4

Real Numbers on an Aristotelian Continuum

4.1 Introduction

Chapter 2 develops a "gunky", or "regions based", account of a one-dimensional continuum. As noted, the account is Aristotelian in one sense: the space is not composed of points. Each part of the continuum has a proper part. We show how "points" (or surrogates for those) can be defined on such structures, in terms of "convergent" sets or sequences of regions, via something like Whiteheadian "extensive abstraction" ([128], ch. 7). Even so, there is no sense in which "points", so construed, are in, or are part of, regions. Indeed, "points" just *are* equivalence classes of sets or sequences of regions. We show that the "points", so defined, are isomorphic to the real numbers, and thus to the so-called "classical" point–set continuum of Cantor and Dedekind.

From the other direction, in a standard Dedekind–Cantor continuum, the regular open sets constitute a model of our theory. The regular closed sets constitute another such model. So, in a sense, our regions-based account is equivalent to the now standard Dedekind–Cantor one, a theory that does, of course, construe a continuum as composed of points (and nothing but points).

As emphasized in the previous chapter, the regions-based theory in Chapter 2 does not follow Aristotle in another, important aspect. We make essential use of so-called "actual infinity". Recall that Aristotle, along with just about every major mathematician and philosopher before the nineteenth century, rejected that notion, insisting that the only sensible concept is that of potential infinity. In *Physics* 3.6 (206a27–9), Aristotle wrote, "For generally, the infinite has this mode of existence: one thing is always being taken after another, and each thing that is taken is always finite, but always different". We thus dub our original, regions-based theory *semi-Aristotelian*. We will show here that this difference with Aristotle is responsible for the fact that our basic, regions-based theory of Chapter 2 is equivalent to its punctiform counterpart.

Chapter 3 develops a more fully Aristotelian account of a one-dimensional continuum. The axiom that allows the use of actual infinity in the semi-Aristotelian framework is our mereological principle of unrestricted fusion, stating that any regions (no matter how many) have a fusion. The first step toward an Aristotelian theory is, in effect, to restrict this to finite collections (or pluralities, or properties) of regions. It turns out, however, that in the original, semi-Aristotelian theory, the use of actual infinity—through unrestricted fusion—allows us to prove several theorems that are, in fact, legitimate (and, indeed, necessary) for the Aristotelian. So in the Aristotelian theory, analogues of several key theorems of Chapter 2 are added as new *axioms*. This at least suggests the fruitfulness of the notion of actual infinity. We also present a second theory that captures the "potential" nature of Aristotelian continua in a modal language (following the main theme of [79] and [81]).

In this chapter, we show how to introduce "points", or, better, "real numbers" into the Aristotelian framework, as a sort of superstructure, along similar lines to how that is done in Chapter 2, via "extensive abstraction". This is an effort to introduce, and develop a *theory of* potentially infinite sequences. It turns out that, unlike the foregoing semi-Aristotelian accounts, the punctiform "superstructure" here is structurally very different from the underlying, gunky space. We take that to be significant, showing the fruitfulness of the actual infinite in unifying several ostensibly different conceptions of the same notion.

To save the reader from having to page back, we briefly recapitulate the basics of the semi-Aristotelian theory of Chapter 2, the introduction of "points" and "real numbers" in that system, via "extensive abstraction", and the first (non-modal) Aristotelian theory of Chapter 3. We only give axioms, definitions, and key theorems.

4.2 A "Semi-Aristotelian" Continuum: The Basics

The background is classical, first-order logic with identity supplemented with a standard axiom system for second-order logic (or a logic of plural quantification, with an unrestricted comprehension axiom for plurals).

1a. Axioms on $x \leq y$ ("x is part of y"): reflexive, anti-symmetric, transitive.

As is standard, we define a binary relation called "overlaps": $x \circ y \Leftrightarrow^{df} \exists z(z \leq x \,\&\, z \leq y)$. And we write $x|y$ for $\neg x \circ y$, pronounced "x is discrete from y".

1b. Axiom on $\leq$ **and** $\circ$: $x \leq y \leftrightarrow \forall z[z \circ x \rightarrow z \circ y]$.

This implies an extensionality principle, yielding so-called classical mereology:

Theorem 1. $x = y \leftrightarrow \forall z[z \circ x \leftrightarrow z \circ y]$.

2. Axiom of Fusion or Whole Comprehension: $\exists u\Phi(u) \rightarrow [\exists x \forall y\{y \circ x \leftrightarrow \exists z(\Phi(z) \,\&\, z \circ y)\}]$, where Φ is a predicate of the second-order language (or language of plurals) lacking free x.

This is the axiom that is dropped in the more Aristotelian theory.

We write $x + y$ for the mereological sum or fusion of x and y. If $x \circ y$, then we write $x \wedge y$ for the *meet* of x and y. It is the fusion of *all* regions that are part of both x and y. So $\forall z[z \leq x \wedge y \leftrightarrow z \leq x \,\&\, z \leq y]$ (and if x and y have no common part, then $x \wedge y$ is undefined).

Similarly, if $\exists z(z \circ x \,\&\, \neg(z \circ y))$, then we let $x - y$ be the sum of all regions z that are part of x but discrete from y (and if there is no such z, then $x - y$ is undefined). So $\forall z(z \leq x - y \leftrightarrow (z \leq x \,\&\, \neg(z \circ y)))$. Again, we do not check to see how many regions are so fused.

We let G be the fusion of *all* regions. It is the entire line, as a single region.

We introduce a geometric primitive, $L(x, y)$, to mean "x is (entirely) to the left of y". The axioms for L specify that it is irreflexive, asymmetric, and transitive. And we define '$R(x, y)$', "x is (entirely) to the right of y", as $L(y, x)$. We adopt the following axioms:

3a. $L(x, y) \vee R(x, y) \rightarrow x|y$.
3b. $L(x, y) \leftrightarrow \forall z, u[\, z \leq x \,\&\, u \leq y \rightarrow L(z, u)]$.

We then define *betweenness*: $Betw(x, y, z)$ for "y is (entirely) between x and z":

$$Betw(x, y, z) \Leftrightarrow^{df} [L(x, y) \,\&\, R(z, y)] \;\vee\; [R(x, y) \,\&\, L(z, y)]$$

It follows that $Betw(x, y, z) \leftrightarrow Betw(z, y, x)$. The following can also be inferred:

$Betw(x, y, z) \rightarrow x|y \,\&\, y|z \,\&\, x|z$, and

$Betw(x, y, z) \,\&\, Betw(u, x, z) \rightarrow Betw(u, y, z)$.

We next define the notion of a *connected* region, one that has no gaps:

$$Conn(x) \Leftrightarrow^{df} \forall y, z, u[z, u \leq x \,\&\, Betw(z, y, u) \rightarrow y \leq x]. \qquad \text{(Df } Conn\text{)}$$

We define what it means for a region to be *bounded* (on both sides):

$$Bounded(p) \Leftrightarrow^{df} \exists x, y\, Betw(x, p, y). \qquad \text{(Df } Bounded\text{)}$$

We call bounded, connected regions "*intervals*" and write '$Int(j)$', etc., when needed.

Using L, we can impose a condition of dichotomy for discrete intervals:

4. Dichotomy axiom: $\forall i, j[i, j$ are two discrete intervals $\rightarrow (L(i, j) \vee L(j, i))]$.

Next is a linearity condition among intervals:

Theorem 2. *(Linearity): Let x, y, z be any three pairwise discrete intervals; then exactly one of x, y, z is between the other two.*

The key Aristotelian feature is the following "gunkiness" axiom:

5. $\forall x \exists j[Int(j) \,\&\, j < x]$.

An important relation of two intervals is "adjacency", which is defined as follows:

$$Adj(j, k) \Leftrightarrow^{df} j|k \,\&\, \nexists m[Betw(j, m, k)]. \qquad \text{(Df } Adjacent\text{)}$$

Intuitively, adjacent intervals touch, but do not overlap.

The following equivalence relation is useful: "*j and k are left-end-equivalent*" just in case

$$\exists p[p \leq j \,\&\, p \leq k \,\&\, \nexists q(\{q \leq j \vee q \leq k\} \,\&\, L(q, p))].$$

"*Right-end-equivalent*" is defined analogously.

Our final geometric primitive is *congruence*, a binary relation among intervals. We adopt the usual first-order axioms specifying that *Cong* is an equivalence relation. We will sometimes write this as $|i| = |j|$. Similarly, for intervals i, j, we can define, contextually, $|i| < |j|$ as meaning: $\exists j'[j'$ an interval $\&\ j' < j\ \&\ \text{Cong}(i, j')]$; and we may write $|i| > |j|$ as equivalent to $|j| < |i|$. Say that $|i| \leq |j|$ just in case either $|i| < |j|$ or $|i| = |j|$.

The next axiom is crucial:

6. Translation axiom: Given any two intervals, i and j, each is congruent both to a unique left-end-equivalent and to a unique right-end-equivalent of the other.

Lemma 1. *Given any two intervals i and j such that* $\neg Cong(i, j)$*, either there exists an interval* $i' < j$ *with* $Cong(i, i')$*; or there exists* i' *with* $j < i'$ *with* $Cong(i, i')$*.*

Theorem 3. *(Trichotomy) For any two intervals,* i, j*, either* $|i| = |j|$ *or* $|i| < |j|$ *or* $|i| > |j|$.

Our final axiom is that congruence respects nominalistic summation of adjacent intervals:

7. Additivity: Given intervals i, j, i', j' such that $Adj(i, j)$, $Adj(i', j')$, $Cong(i, i')$, $Cong(j, j')$, then $Cong(k, k')$, where $k = i + j$ and $k' = i' + j'$.

A key theorem:

Theorem 4. *(Bi-Infinity of G) Let any interval i be given; then there exist exactly two intervals, j, k, such that* $Cong(i, j)$ *&* $Cong(i, k)$ *&* $Adj(i, j)$ *&* $Adj(i, k)$*, and one of j, k is left of i and the other is right of i.*

Call an interval l an (*immediate*) *biextension* of interval i—$BiExt(l, i)$, or $biext(i) = l$—just in case $l = j + i + k$, where j, i, k behave as in the Bi-Infinity theorem.

Lemma 2. *Let i and j be intervals such that* $i < j$*; then* $\neg Cong(i, j)$.

One central result is the Archimedean property that, in effect, our entire space G is *exhausted* by iterating the process of flanking a given interval by two congruent ones as in Bi-Infinity.

Let X be any class (or plurality) of intervals such that an arbitrary but fixed interval $i \leq G$ is one of the X and such that if $k = biext(j)$ for j any of the intervals of X, then k is also in X. Call such X a "closure of i under *biext*".

By axiom 2, there is an individual which is the common part of the fusions of each class X which is a closure of i under *biext*, which we call their *meet* or *the minimal* *closure i^* of i under biext.* (Since i is stipulated to belong to any such X, the meet is non-null, as required in mereology.)

Theorem 5. *(Archimedean property): Let G be the fusion of the objects in the range of the quantifiers of our axioms; then $G = i^*$, the fusion of the minimal closure of i under biext.*

We then establish that any interval has a unique bisection:

Theorem 6. *(Existence and uniqueness of bisections): Given any interval i, there exist intervals j, k such that $j < i \,\&\, k < i \,\&\, j|k \,\&\, j + k = i \,\&\, Cong(j, k)$; and j, k are unique with these properties.*

4.3 Semi-Aristotelian Superstructure: Putting "Points" Back in

Chapter 2 provides a number of different ways to define "points" and "real numbers" in this semi-Aristotelian, gunky space, as a sort of set-theoretic superstructure—again following Whiteheadian "extensive abstraction". As in section 2.4, define a set I of intervals to be *convergent* if (i) if i and j are in I, then either $i \leq j$ or $j \leq i$, (ii) if i and j are in I, and if $i \neq j$, then i is neither right-end or left-end-equivalent to j, and (iii) there is no region that is part of every member of I. Intuitively, a set of intervals is convergent if its members surround a single location. We then define an equivalence relation on convergent sets of intervals: I is equivalent to J just in case every member of I overlaps every member of J. Then a "point" is an equivalence class of convergent sets of intervals. It follows that the "points" so defined, are isomorphic to the real numbers.

Although we do not work out the details, one can also define a "point" to be an equivalence class of intervals under the relation of left-end-equivalence (or right-end-equivalence). The "points", so construed, also have the structure of the real numbers.[1] One can also define a "length" to be an equivalence class of intervals under congruence. The "lengths", so defined, are isomorphic to the positive real numbers.

Our main recovery of the real numbers (or something that can play the role of real numbers), in section 2.2, goes via the proof of the bisection property (Theorem 6) and Axiom 2 of fusions. Rather than go back over the details, we illustrate the method with an example.

Let i be any interval, which we regard as a unit, standing for the real number 1. Let a_0 be the fusion of i with an interval congruent to i adjacent on its right, and another interval congruent to i adjacent to that on *its* right. The idea is that a_0 corresponds to

[1] In [111] we show how to recover "points" in this framework (and the more Aristotelian one) as neo-logicist abstraction principles.

the real number 3. Next consider the fractional part of π as a bicimal expansion. It starts 0010010000111111 . . . (see http://www.befria.nu/elias/pi/binpi.html). Define a sequence of intervals, beginning with a_0 as follows:

If the nth member of the bicimal expansion of π is 0, then $a_{n+1} = a_n$; and
if the nth member of the bicimal expansion of π is 1, then a_{n+1} is the fusion of a_n with an interval congruent to an n-fold bisection of i, adjacent to a_n and on its right.

The real number π corresponds to the fusion of the members of this sequence $\langle a_n \rangle$.

The positive real numbers thus correspond to intervals that are left-end-equivalent with the given unit i. It is then straightforward to define negative real numbers as intervals that are adjacent to i on the left (where, for each positive real number r, the interval corresponding to $-r$ is congruent to the interval that corresponds to r). Then special provision is given for the number 0.

Notice that each real number thus corresponds to a *single* interval and thus to a *region* in the gunky line G. One can show that the set of indicated intervals is indeed isomorphic to the real numbers.

Thus we have another way to recapture the now standard Dedekind–Cantor continuum *within* our gunky line (in some sense of "within"). In light of the topological models, mentioned above (see section 2.3 for details), our semi-Aristotelian gunky theory is equivalent to the now standard, Dedekind–Cantor punctiform theory. It is easy to go back and forth between the frameworks.

It is not entirely clear, at least not to us, just what it is to treat a given set or sequence as an actual infinity, as opposed to taking it to be merely potential. More on this below. It seems clear, however, that each of the reconstructions of the real numbers in our semi-Aristotelian space invokes a clear case of actual infinity. In finding the analogue for the number π, for example, we take the fusion of an infinite sequence $\langle a_n \rangle$ of intervals. That is, we envision a *single* region—an interval—that fuses *all* of the members of the sequence. To envision this, we have to think of the sequence as somehow given in its entirety, as "finished". We cannot be thinking of the sequence as somehow forever in progress.

4.4 Basic Aristotelian Theory

As noted, the semi-Aristotelian treatment invokes infinity in two related ways, neither of which is acceptable to an Aristotelian. First, there are regions, such as the entire Gunky line G, that are infinitely long (relative to any interval). The second and more substantial difference is that in the proofs of some of our theorems, we take fusions of infinite sets (or pluralities or properties) of regions. For an Aristotelian, there is no conceptual problem with there being a *potential* infinity of regions. But, again, one cannot take the fusion of infinitely many such regions—to produce a single region—unless we take the infinity to be "actual".

The play with infinity is sanctioned by the above Axiom 2, the principle of fusions or whole comprehension:

$$\exists u \Phi(u) \to [\exists x \forall y \{y \circ x \leftrightarrow \exists z(\Phi(z) \,\&\, z \circ y)\}],$$

where Φ is a predicate of the second-order language (or language of plurals) lacking free x.

A natural Aristotelian "fix" to our system would be to replace this with an axiom asserting the existence of fusions of finite sets. For this, of course, it is sufficient to state that any two regions have a fusion:[2]

$$\forall u \forall v \exists x \forall y [y \circ x \leftrightarrow (y \circ u) \vee (y \circ v)]. \qquad \text{(FINFUS)}$$

Let A^0 be the theory that results from our semi-Aristotelian theory by removing the axiom (2) for unrestricted fusion and adding (FINFUS).

Unfortunately, A^0 is too weak. It cannot establish the existence of the meet of overlapping intervals, nor the existence bisections, biextensions and appropriate differences. We presume that those are all acceptable to an Aristotelian, and, indeed, go with the relevant conception of space.

Let A^1 be the result of adding to A^0 an axiom of *differences*:

If $\exists z(z \circ x \,\&\, \neg(z \circ y))$ then $\exists w \forall z(z \leq w \longleftrightarrow (z \leq x \,\&\, \neg(z \circ y)))$.

Again, if $\exists z(z \circ x \,\&\, \neg(z \circ y))$ then let $x - y$ be the resulting difference.

Theorem A1. A^1 *implies the existence of biextensions of intervals and the meet of any overlapping regions.*

We still do not have bisections, or n-sections generally. So we add a principle asserting that, for each positive integer n, there are n-sections:

(n-SECT) Given any interval i, there exist intervals $j_1, \ldots, j_n$, all congruent to each other, all part of i, and such that for each $m, m' \leq n$, if $m \neq m'$, then j_m is discrete from $j_{m'}$, and $i = j_1 + \ldots + j_n$.

One could just take this as an axiom scheme, one instance for each natural number n. It would perhaps be better to add terminology for the natural numbers to the system, with appropriate axioms. In the Aristotelian spirit, the natural numbers would be thought of as a potential infinity. Let A^2 be the resulting theory, A^1 plus (n-SECT).

An Archimedean property still does not follow from A^2. In fact, we adopt an Archimedean property that is stronger than the one appropriate to the semi-Aristotelian system.

[2] We also propose an alternative but equivalent fix. The idea is to "factor" (FINFUS) into components: two principles licensing the formation of any *finite* plurality; and, then, a principle stating that any plurality has a fusion. The first such principle states the existence of all "singleton pluralities" and the second states that any object may be added to any plurality to yield a plurality.

If i is an interval and n any natural number greater than zero, define the n-fold biextension of i, written ni, by recursion, as follows:

$1i = i$
$(n + 1)i$ is the fusion of ni with an interval congruent with i and adjacent on its left and an interval congruent with i adjacent on its right.

Our main Archimedean principle is this:

(ARCH2) Let i be an interval and let k be any region. Then there is a natural number n such that $k < ni$.

Of course, the semi-Aristotelian theory *refutes* (ARCH2), since the theory proves the existence of infinitely large regions, such as the entire space G. However, we think that an Aristotelian would endorse this stronger Archimedean principle.

Let A^3 be the theory obtained from A^2 by adding (ARCH2). This our first full Aristotelian theory.[3]

We now turn to the recovery of "points" and "real numbers" in this theory.

4.5 Superstructure: Putting "Points" Back in (and Taking Potential Infinity Seriously)

In both the above semi-Aristotelian and Aristotelian spaces, the first-order variables range over (extended) regions. Parts of regions are other regions, and every region has an interval as a (proper) part. So (once again) neither of those theories, nor any of their variants, recognize points or point-sized regions, at least as parts of regions. As noted, in the original, semi-Aristotelian theory, we are able to define "points" and even "real numbers", as a kind of superstructure.

It is clear, however, that the procedures for recapturing the Dedekind–Cantor punctiform continuum, "extensive abstraction", are rather unAristotelian. The interval that corresponds to an irrational number, such as π, is a fusion of an infinite set of regions. An Aristotelian might be able to make sense of our definition of the above *sequence* $\langle a_n \rangle$ of intervals, as a potential infinity. In effect, we define a *procedure* for producing more and more of the intervals a_n, a procedure that can be carried out as long as one wishes, without limit. In this case, the procedure is fully effective—a routine recursive function. From the Aristotelian perspective, the sequence itself is not that much different from the procedure of starting with an interval and continually taking the left bisect, a standard Euclidean construction—one that can be executed, effectively, as far as one wishes. But surely we crossed the line when we concluded (via

[3] Chapter 3 also contains a theory that attempts to capture the "potential" nature of Aristotelian construction (and infinity) in a modal language. For example, instead of an axiom that any interval has an n-section, the modalized theory has an axiom stating that, necessarily, any interval *can be* n-sected. Following the theme of [79] it is shown that the modalized theory is equivalent to A^3.

the unrestricted Axiom 2, of fusions) that there is a *single* region, within the gunky space G, that is the fusion of *all* of the members of the sequence $\langle a_n \rangle$. If we think of the sequence as a *process*, we have to somehow finish it before we can take its fusion.

In the contemporary philosophical/mathematical scene, one can, we suppose, think of a sequence like $\langle a_n \rangle$ as a procedure, and then allude to the so-called "super-task" of finishing it. The transfinite ordinals might provide a good backdrop for this bit of theorizing. But that way is fraught with conceptual problems, and surely *that* presupposes actual infinity. Better to avoid talk of super-tasks.

To be sure, the contemporary mathematician does not think of the sequence $\langle a_n \rangle$ as a procedure at all. It is an (actually) infinite *set* of ordered pairs. We simply take the fusion (or union) of the intervals that appear in this set.

The plan here is to mimic, as far as is possible, the superstructure recapture of the real numbers, working from the (unmodalized) one-dimensional Aristotelian continuum presented in Chapter 3, and summarized just above. We define a real number to be an equivalence class of Cauchy sequences of regions, with each Cauchy sequence itself construed as a (merely) potential infinity.[4]

Admittedly, we have to go beyond Aristotelian resources for this task, but not in the direction of contemporary so-called "classical" mathematics. We remind the gentle reader that we are not (even close to) experts in the history of mathematics and philosophy. That said, it is our impression that the attitude toward infinity from antiquity through the modern period was mostly negative. Most of the great mathematicians and philosophers rejected the existence of actually infinite collections, entities, or whatever.

Throughout this history, *particular* potentially infinite processes were envisioned, and put to great, indeed brilliant use. Consider, for example, Euclid's (Book 10, Proposition 1) formulation of an Archimedean property, cited above:

> Two unequal magnitudes being set out, if from the greater there be subtracted a magnitude greater than its half, and from that which is left a magnitude greater than its half, and if this process is repeated continually, there will be left some magnitude less than the lesser magnitude set out.

The ancient method of exhaustion for, among other things, determining the areas of curved figures, depends on this principle.

What the ancient and early modern thinkers did not do, as far as we know, is develop a *theory of* potentially infinite sequences or processes. To put it a bit anachronistically, the mathematicians and philosophers in question did not envisage variables that *range over* such processes or sequences, at least not with the proper depth and level of rigor we (sometimes) have today. The various schools of intuitionism and

[4] In the meta-theory of the modalized version of the Aristotelian theory, a "Cauchy sequence" of regions corresponds to something like a path through the space of possible worlds. Officially, however, we do not take the talk of possible worlds literally. It is only heuristic. There seems to be no convenient way to talk, in the object language, of paths through possible worlds.

constructive mathematics, through the twentieth century, did just that. Of course, matters of logic get bound up with such investigations.

With a little more detail, the present plan is to show, first, how to embed the *rational* numbers into the (unmodalized) Aristotelian continuum. That part is, we think, straightforward, and does not go beyond Aristotelian resources. Once a unit is fixed, each rational number corresponds to an interval in the space. Next, we define, in terms of our structure, what it is for a sequence of such rational-number-surrogates to be Cauchy, and what it is for two such Cauchy sequences to be equivalent. Then a "real number" will be defined to be an equivalence class of Cauchy sequences of such "rational" intervals.

So, unlike the analogous case with the original "semi-Aristotelian" theory, a "real number" here does *not* correspond to a region in the space (although each rational number does correspond to an interval). As above, we cannot simply take the fusion of the members of an infinite sequence—not if we insist that each such sequence is merely potentially infinite.

Crucially, the "real numbers", so construed, are very different in structure from the underlying gunky space (and also different from the Dedekind–Cantor real numbers). We have, for example, that, in the gunky space, each interval can be bisected cleanly. An instance of our axiom (n-SECT) tells us that if i is an interval, then there are discrete and congruent intervals j and k such that i is the fusion of j and k. So our gunky, Aristotelian intervals are decomposable. We will not be able to prove a similar property concerning the set of "real numbers" that, intuitively, corresponds to an interval i. Indeed, given some natural theoretical choices to be made along the way, we can actually show that such "intervals" *cannot* be decomposed in the indicated manner. This is a well-understood feature of intuitionistic and constructive continua.

As suggested above, our major conceptual issue here lies in how to understand variables that themselves range over potentially infinite sequences (or sets). Our theory of real numbers—or real-number-surrogates—requires this. Intuitively, we have somehow to talk about *all* such sequences (or all such sequences that have a certain property), insisting that *each* such sequence is only potentially infinite. The key feature, we think, is that any conclusion one can draw about a given sequence s has to be based on a finite amount of information I about s. So if this same information I held for another sequence s', then the same conclusion would apply to s'. The underlying theme is that we "never" have the entire sequence s available. It is, after all, only potentially infinite. Typically, a finite bit of information I about a given sequence s is either an initial segment of s or a rule for generating the remaining elements of s, or some combination of those two things. The key to the theory is to exploit that limitation.

As noted, the contemporary, Cantorian set-theoretic approach to sets and sequences is not appropriate here. There, each sequence is simply treated as an actual infinity. Of course, we are not claiming that there is something wrong with treating sequences this way, only that it is not appropriate here, in trying to capture the notion of potential infinity.

As noted just above, better models for us are found in contemporary constructive mathematics. Constructivists, however, do not speak with one voice on these matters. Michael Beeson [12] identifies at least six schools of constructivism. For present purposes, the three main ones are intuitionism (L. E. J. Brouwer and Arend Heyting), Russian constructivism (led by A. A. Markov), and Bishop–Bridges constructivism (which is consistent with so-called "classical" analysis and also with intuitionism and Russian constructivism). We will try to stay neutral on this, as far as possible. However, all of these theories do seem to require a non-classical logic, at least for the real numbers. We will see how far this restriction reaches.

4.6 Rational Numbers

We work in our theory A^3, which presupposes the natural numbers (themselves understood as a potentially infinite set).[5] As sketched just above, our reconstruction of the real numbers in the original "semi-Aristotelian" theory, begins with an arbitrary interval i, which is identified as the number 1. Each positive real number corresponds to an interval that is left-end-equivalent with i. If $r < 0$, then r corresponds to an interval that is congruent to the interval corresponding to $-r$, and adjacent to the unit interval i on its left. A special provision is made for the number 0 (since we do not recognize "empty" regions).

The embedding in the semi-Aristotelian theory thus makes essential use of the law of excluded middle. In particular, we assume that each real number is either greater than, identical to, or less than zero. Of course, that was not problematic there (or so we contend or at least assume), since our target was the Dedekind–Cantor theory of real numbers, which assumes classical logic (and the present authors have no problem with that). However, we do not want to presuppose classical logic here, at least not yet, so that we can remain neutral on certain issues to emerge soon. So the present construction begins in a somewhat different way.

Let i be any interval, chosen arbitrarily. That is to correspond to the rational number 0. Each rational number is to correspond to an interval that is congruent with i. Define a function from natural numbers to intervals (congruent with i), as follows:

$i(0) = i$.

$i(n + 1)$ is the interval congruent with i and adjacent to $i(n)$ on its right.

This gives us analogues of the natural numbers.

Now let n be any natural number. $i(-0)$ is, of course, $i(0)$, which is i. We define:

$i(-(n + 1))$ to be the interval congruent with i and adjacent to $i(-n)$ on its left.

[5] It would be straightforward to "translate" the contents of this section into the Aristotelian framework that invokes (finite) plurals instead of natural numbers, and also into the modalized theory.

The idea is, of course, straightforward. We start with i, identified with 0, and keep taking intervals congruent to i, to its left and right. Think of this as a potentially infinite set. So we have surrogates for the integers.

Recall that for each positive integer n, we have a principle of n-sections:

(n-SECT) Given any interval i, there exist intervals $j_1, \ldots, j_n$, all congruent to each other, all part of i, and such that for each $m, m' \leq n$, if $m \neq m'$, then j_m is discrete from $j_{m'}$, and $i = j_1 + \ldots + j_n$.

If k is an interval and m is a positive integer, then let the m m-sections of k be $k_0^m, k_1^m, \ldots, k_{m-1}^m$, where k_0^m is the left-most (so that k_0^m is left-end-equivalent to k), k_1^m is adjacent to k_0^m and right of k_0^m, and so on. The last sub-interval k_{m-1}^m is right-end equivalent with k.[6]

We define a function f from rational numbers r to intervals congruent with i, as follows. If $r \geq 0$ then let $[r]$ be the greatest integer that is less than or equal to r, and let m, n be natural numbers such that $r = [r] + \frac{n}{m}$ (so that $n < m$). The interval $f([r])$ that corresponds to $[r]$ is, of course, $i([r])$. The interval $f(r)$ that corresponds to r is congruent to i and left-end-equivalent with $i([r]_n^m)$, the nth m-section of $i([r])$.

From the standard properties of the natural numbers, it is routine to check that if r is itself a natural number, then, since $[r] = r$, $f(r)$ is the interval $i(r)$. And if $\frac{n_1}{m_1} = \frac{n_2}{m_2}$, then $i([r]_{n1}^{m_1})$ is left-end-equivalent with, and thus congruent to, $i([r]_{n_2}^{m_2})$. So our definition is well-defined (and the fraction does not have to be given in lowest terms).

The analogues of negative rational numbers are the same, but from the opposite direction. So let $r > 0$, and, as above, let $r = [r] + \frac{n}{m}$, with $n < m$. If $n = 0$ (and thus r is a natural number), then $f(-r)$ is the interval $i(-[r])$ (i.e. $i(-r)$). If $n \neq 0$, then $f(-r)$ is the interval that is congruent with i and left-end-equivalent with $i(-([r]+1)_{m-n}^m)$.

Say that an interval j is a *rational-number-analogue* if there is a rational number r such that $f(r) = j$. Intuitively, r is a sort of measure of the left-end of $f(r)$. So, for example, the left end of $f(\frac{1}{3})$ is $\frac{1}{3}$ of the length of i to the right of the left end of i. But, of course, this talk of "ends" is only a manner-of-speaking.

It is straightforward to define the order relation on the rational-number-analogues: Say that $f(r) < f(s)$ just in case there is a part of $f(s)$ that is right of $f(r)$, or, equivalently, if there is a part of $f(r)$ that is left of $f(s)$. The axioms of the Aristotelian theory entail that $f(r) < f(s)$ if and only if $r < s$. If we wish, we can define addition and multiplication on the analogues, in terms of the corresponding operations on the rational numbers.

So our Aristotelian space has an isomorphic copy of the rational numbers. So, we do not have to go beyond Aristotelian resources to recapture rational analysis (assuming

6 The procedure here is similar to the embedding of the binary rationals in the gunky continuum in the final section of the previous chapter, on predicativity.

that it is acceptable to presuppose the natural numbers, or something equivalent, such as finite pluralities, in formulating the theories).

4.7 The Matter of Logic—So Far

It may be too much of an anachronism to attribute a contemporary, sophisticated logic—classical, intuitionistic, relevant—to Aristotle, but we do need to make some decisions for the present treatment. Our original semi-Aristotelian treatment, above, invokes ordinary, so-called "classical" logic, but there we were not attempting to be faithful to Aristotle, even by way of rational reconstruction. That's why we called it *semi*-Aristotelian. In fact, one of our goals was to recapture a version of the now orthodox Dedekind–Cantor picture of continua in a point-free setting. As noted, we made free play with (actual) infinity. Our primary question here, where we are trying to stay closer to Aristotelean resources, is whether we can safely assume the principle of excluded middle, and adopt classical logic (putting aside any issues of relevance logic).

The principle of excluded middle is one of Aristotle's first principles, and at least in serious scientific and metaphysical discourse, he does not seem to allow for exceptions to it. In *Metaphysics* 4.7, he writes:

> [T]here cannot be an intermediate between contradictories, but of one subject we must either affirm or deny any one predicate. This is clear, in the first place, if we define what the true and the false are. To say of what is that it is not, or of what is not that it is, is false, while to say of what is that it is, and of what is not that it is not, is true; so that he who says of anything that it is, or that it is not, will say either what is true or what is false.

Aristotle does not here distinguish excluded middle from the related semantic principle of bivalence.[7] His target may be views that postulate an intermediate truth-value: "While the doctrine of Heraclitus, that all things are and are not, seems to make everything true, that of Anaxagoras, that there is an intermediate between the terms of a contradiction, seems to make everything false" (1012a25–9).

The opening of chapter 9 of *De Interpretatione* endorses bivalence for propositions about the present and propositions about the past: "In the case of that which is or which has taken place, propositions, whether positive or negative, must be true or false." However, "When the subject . . . is individual, and that which is predicated of it relates to the future, the case is altered." One of the most discussed, and most puzzling, passages in the Aristotelian corpus concerns so-called future contingents:

[7] In treatments of contemporary intuitionism, the phrase "law of excluded middle" is something of a misnomer. The intuitionist does "exclude the middle" in that she denies that a proposition can be something other than true or not true (again putting aside the distinction between bivalence and excluded middle). In the passage above, Aristotle seems to make the now-disputed inference from "lack of a middle" to the law of excluded middle, and also, it seems, to bivalence. Thanks to Sarah Broadie for help in understanding this material.

It is necessary for there to be or not to be a sea-battle tomorrow; but it is not necessary for a sea-battle to take place tomorrow, nor for one not to take place—though it is necessary for one to take place or not to take place. So, since statements are true according to how the actual things are, it is clear that wherever these are such as to allow of contraries as chance has it, the same necessarily holds for the contradictories also. This happens with things that are not always so or are not always not so. With these it is necessary for one or the other of the contradictories to be true or false—not, however, this one or that one, but as chance has it; or for one to be true rather than the other, yet not already true or false. Clearly, then it is not necessary that of every affirmation and opposite negation one should be true and the other false. For what holds for things that are does not hold for things that are not but may possibly be or not be; with these it is as we have said. (19a30–b4)

To us, this passage defies interpretation. In various places, it has been taken to endorse a kind of fatalism, supervaluation, relativism, and many other positions in the philosophy of language, logic, metaphysics and the philosophy of time. We will stay neutral on as much of that as possible. Notice that Aristotle does insist that it *is* "necessary for there to be or not to be a sea-battle tomorrow". So he does seem to endorse excluded middle, even for future contingents.

As above, we assume that our main options for the underlying logic of the various theories are intuitionistic logic and (so-called) classical logic, or perhaps some intermediate logic. For the natural and rational numbers, the issue is not central, or not as central. The reason is that the intuitionist agrees that, for those domains, identity is decidable. Using the induction principle, along with the usual properties of the natural numbers, the intuitionist easily proves:

$$(\forall m)(\forall n)(m = n \vee m \neq n)$$

We also have that, for every rational number r, either $r > 0$, $r = 0$, or $r < 0$. So the intuitionist and the classicist agree on that much. This sanctions the above, piecemeal definition of the embedding of the rational numbers in the Aristotelian space. We do not think we have stepped on any intuitionistic toes, at least not yet.

But, of course, excluded middle does matter elsewhere in the theories of natural and rational numbers. The more challenging questions include the so-called Brouwerian weak counterexamples, which are, in effect, Π^0_1 statements. For example, let P be the statement of the Goldbach conjecture or, to use an example like one of Brouwer's, let Q be the statement that the decimal expansion of π contains a sequence 0123456789 somewhere. Can we (now) assert the statements $P \vee \neg P$ and $Q \vee \neg Q$?

There is nothing contingent about the various instances of the Goldbach conjecture nor about the decimal expansion of π. So this is not a sort of sea-battle case (unless fatalism is true of sea-battles). In any case, a better analogue would be a statement that there will *sometime* be a sea-battle in a given location. Perhaps we can cite Aristotle's assertion of the instance of excluded middle for his own sea-battle statement, along with the fact that he seems to allow no exceptions to excluded middle. Maybe.

We seem to be asking something about the *entire* sequence of natural numbers, namely whether there is a member of it that has a certain (effectively decidable) property—either being an even number greater than 2 that is not the sum of two primes, or being a first place in the decimal expansion of π that ends 0123456789. Can we assert these instances of excluded middle, and still regard the sequence of natural numbers as only potentially infinite?[8] Almost all constructivists say "no".

However, potential infinity and constructivity are not the same (or might not be the same). We will stay neutral here. It will, however, not be so easy to remain neutral on the corresponding issues when we look to recapture the real numbers in our Aristotelian space. Stay tuned.

4.8 Cauchy Sequences

Say that an interval j *left-overlaps* an interval k if $j \circ k$ and a part of j is left of k. In other words, j left-overlaps k just in case j has a part that overlaps k and another part that lies entirely to the left of k. So a bit of j sticks out on the left side of k.

In the Aristotelian theory developed above, we have resources to speak of regions and (finite) pluralities of regions. We now introduce *sequences* of intervals as a new primitive in the system. If terminology for the natural numbers is available, then one can think of a sequence as a function whose domain is the natural numbers. The goal here is to treat these sequences (or functions) as (merely) potential infinities.

Once again, it may well be acceptable for an Aristotelian to speak of *individual* sequences, say as specific processes of a certain kind. As above, however, we do think we are going beyond Aristotelian resources in developing a theory with bound variables ranging over such sequences.

Say that a sequence $a = \langle a_0, a_1, \ldots \rangle$ of intervals is *left-Cauchy* if for any interval j there is an interval j', congruent with j and a natural number N, such that for any $m > N$, j' left-overlaps a_m. In other words, a sequence a is left-Cauchy just in case for any interval j, no matter how small, eventually the "left-ends" of the members of a all lie in an interval congruent to j.

It will be convenient to refine this a bit. Recall that i is the fixed, but arbitrary interval used to recapture the rational numbers. Let $a = \langle a_0, a_1, \ldots \rangle$ be a sequence of rational-number-analogues (i.e. certain intervals congruent with i). Say that a sequence a is a *real-number-generator* if for every positive integer n, there is an interval j_n congruent to an n-section of the given unit interval i, and left-end-equivalent with a rational-number-surrogate, such that for any $m > n$, j_n left-overlaps a_m.

Intuitively, the idea is that for any positive integer n, after n iterations, the left-ends of the sequence are all within $1/n$ of the unit interval of each other. It follows from the Archimedean property that every real-number-generator is left-Cauchy.

[8] Of course, the issue gets even more vexed if we go beyond these Π^0_1 (and Σ^0_1) statements.

Let $a = \langle a_0, a_1, \ldots \rangle$ and $b = \langle b_0, b_1, \ldots \rangle$ be real-number-generators. Say that a is *equivalent* to b, written $a \simeq b$, if for every positive integer n, there is an interval j_n congruent to an n-section of the given unit interval i and a natural number N, such that for any $m > N$, j left-overlaps *both* a_m and b_m. It is straightforward to verify that this is an equivalence relation. Intuitively, if $a \simeq b$, then the two generators represent the same real number.

One issue that we shall have to face is whether, in the above definitions, the number N and the interval j' have to be obtainable effectively from the given number n. For the time being, let us leave that open.

Define a *real-number-surrogate* to be an equivalence class of real-number-generators. We will sometimes speak of a given real-number-generator a as itself a real-number-surrogate, taking a to represent its equivalence class. It is tedious, but straightforward to check that our definitions are well–defined.

It is also straightforward to define the proper order relation on real-number-surrogates. Let $a = \langle a_0, a_1, \ldots \rangle$ and $b = \langle b_0, b_1, \ldots \rangle$ be real-number-generators (representing real-number-surrogates). Then $a < b$ just in case there are two intervals j, k, such that j is left of k, and there is a natural number N such that for any natural number $m > N$, j left-overlaps a_m and k is left of (and thus discrete from) b_m. Intuitively (and so unofficially), $a < b$ just in case there is an interval j such that, eventually, the left ends of the members of a are all in j, and the members of b are well to the right of that, separated by the interval k. Define $b > a$ if $a < b$.

If the background logic here is constructive, then our definition is a strong one, requiring that, in effect, the real-number-surrogate a be separated from the real-number-surrogate b by an interval k.

Addition, multiplication, exponentiation, etc. can be defined on real-number-surrogates. The definitions would be based on the same operations on the rational numbers (and, ultimately, the natural numbers), through the rational-number-surrogates. That facilitates our recapture of the real numbers.

Recall that for each rational number r, there is an interval $f(r)$, congruent to the unit interval i. The interval $f(r)$ is the surrogate for r. The rational numbers, and their surrogates, can then be embedded in the real-number-surrogates, as follows. If r is a rational number, then its real-number-surrogate $F(r)$ is the (equivalence class of) the sequence $\langle f(r), f(r), \ldots \rangle$. Clearly, this is a real-number-generator, and, of course $r_1 < r_2$ if and only if $f(r_1) < f(r_2)$ if and only if $F(r_1) < F(r_2)$.

4.9 The Matter of Logic (More Serious this Time)

Suppose that in the foregoing development, the background logic is classical. Suppose, also, that in the definitions of left-Cauchy, and the like, we follow contemporary (classical) practice and do not require that the natural number N or the interval j' be given effectively from the relevant inputs. We can then show that for any real-number-surrogates a, b, either $a < b$, $a > b$, or $a = b$ (i.e. for any real-number-generators, either

$a < b$, $a > b$, or $a \simeq b$). Indeed, it is straightforward to check that the surrogate-real-numbers are isomorphic to the now orthodox Dedekind–Cantor real numbers. We have just transposed the more usual Cantorian characterization of the real numbers—as Cauchy sequences of rational numbers—into the present Aristotelian framework, with an extra layer for the surrogate-rational-numbers. Nothing is new or special (or particularly interesting). We would be treating each left-Cauchy sequence and each real-number-surrogate as an *actually infinite* set of intervals, along Dedekind–Cantor lines.

As announced, the orientation of this chapter is to think of the real-number-generators as only *potentially* infinite processes. Admittedly, there is not much to go on in saying what makes for potential infinity, and so we must get a bit speculative.

Once again, our treatment goes beyond Aristotelian resources, even if it remains incompatible with much contemporary, so-called "classical" mathematical practice. As noted, ancient, medieval, and early modern mathematics recognized particular processes, such as that of bisection, that were viewed as potentially infinite, but, as far as we are aware, there was no *theory of* potentially infinite sequences, as such. The constructive mathematicians of the twentieth century and beyond (attempt to) supply just that, but they do not speak with one voice on this. Moreover, this broaches matters of logic. It is perhaps not obvious that "potentiality"—however it is to be developed—is the same as "intuitionistic" or "constructive", especially on matters of logic. In particular, it remains to be seen whether we must invoke a non-classical logic when dealing with potentially infinite sets or sequences. Our goal is to remain as neutral as possible on as many substantial issues as possible.

The main feature of potentially infinite sequences is that they be thought of as *processes* (or as associated with processes). What makes these processes "infinite" is that they can be executed indefinitely, without limit. What makes them merely "potential" is that any information we can have about a given process a has to be based on a finite amount of information about it. The information in question can be either a rule for generating the elements of a or an initial segment of a, or some combination of those.

The matter comes to a head, we think, when one considers *functions* from potentially infinite sequences to potentially infinite sequences. For example, let H be a function from real-number-generators to real-number-generators. Say that H is a *real-valued-function* if, for all real-number-generators a, a', if $a \simeq a'$ then $H(a) \simeq H(a')$.

Let a be a real-number-generator in the domain of a real-valued function H. Then both a and $H(a)$ are potentially infinite sequences of intervals. So any information we have about $H(a)$ has to be based on a finite amount of information—presumably information about a. Suppose, for example, that we want the fourth element of $H(a)$ (so that all subsequent members of the sequence are within $1/4$ of the length of the unit interval i of each other). This is to ask for an approximation to the generated real number. By hypothesis, we have to determine this fourth element of $H(a)$ on the basis of a finite amount of information I of a—we cannot "wait" for the entire sequence a

to be produced before determining the fourth element of $H(a)$. So if that same finite information I held of another sequence a', then the fourth element of $H(a')$ would be the same as the fourth element of $H(a)$.

Our present goal is to explore the thesis that functions from real-number-surrogates to real-number-surrogates, so construed, are all continuous. All, or most, of the main branches of constructivism maintain a continuity principle like this, for rather different reasons in each case (stay tuned). This, of course, is inconsistent with contemporary real analysis, which boasts plenty of discontinuous functions. It also has ramifications for the indecomposability of continua and, indeed, for adjudicating some of the long-standing intuitive features of continua generally (see the closing section of this chapter). Here we will have to break our discussion into pieces, depending on which branch of constructivism we wish to emulate.

4.9.1 *Intuitionism à la choice sequences*

Here is a rather crude argument that all real-valued-functions, as construed here, are continuous. Let H be such a function and let a be in the domain of H. Suppose that we want to estimate the value of $H(a)$ to within, say, n^{-1}, for some positive integer n. That is, suppose we want the nth element of $H(a)$. As noted, this has to be based on a finite amount of information about the sequence a. In general, the finite information in question is an initial segment of a, say the first m elements of it. But that only determines a to within a bound of m^{-1}. So if a' is any other real-number-generator within m^{-1} of a, then $H(a')$ is within n^{-1} of $H(a)$. It follows that H is continuous.

This argument is, of course, too crude. One can obtain finite bits of information about a given real-number-generator other than initial segments of it. In particular, a *rule* for generating the members of a is itself a finite amount of information about a, and that rule might be used to determine some, or even all of the values of $H(a)$. Consider, for example, a constant function, or perhaps the absolute value function, where $G(a) = |a|$.

Brouwer introduced sequences that are not generated by a rule, so-called *lawless sequences*. There we do not have a rule for the sequence to go on. The more interesting specimens of lawless sequences are the so-called choice-sequences. We can adapt those to the present framework. Think of a choice sequence as a real-number-generator, with its elements given in time. At any given moment, it may be genuinely indeterminate what some of the future elements of the sequence are. For example, with a given choice sequence, it may be indeterminate whether it converges to a rational number or an irrational number—in the sense that there are some continuations of the sequence that converge to a rational number and other continuations that converge to an irrational number. And the question of whether the number is rational or irrational may very well remain open no matter how many elements of the choice sequence are determined—no matter how much information we get about the choice sequence. So the question of whether this choice sequence is a rational number is a bit like the above question of whether there will someday be a sea-battle at a given location. That, at least, puts some pressure on bivalence, if not excluded middle (despite Aristotle's seemingly blanket endorsement of excluded middle). We can get sharper here.

Choice sequences are modeled by branching trees. Each branch represents a way the sequence could be continued. Brouwer showed that, from this perspective, it follows that all total, real-valued functions are continuous. Moreover, by adopting what is known as "bar induction", Brouwer established the "fan theorem", a sort of contrapositive to König's lemma. From these theses it follows that all real-valued functions on a bounded interval are uniformly continuous. See Beeson [12], III, §4; Bridges and Richman [26], 5; or Troelstra and van Dalen [120], 12.

With classical logic, of course, it is easy to define discontinuous functions—indeed, discontinuous functions are common fare in classical analysis (and in classical mechanics). Just let $H(a) = 0$, if a converges to a rational number and let $H(a) = 1$ otherwise (with excluded middle sanctioning the "otherwise"). So Brouwerian intuitionistic analysis is inconsistent with classical logic.[9]

However, the use of intuitionistic logic can be attenuated a bit. Applying the techniques of Dana Scott [103], [104], using a classical meta-theory, it can be shown that there are topological models of intuitionistic analysis in which the background first-order arithmetic (of the natural and rational numbers) is classical. The continuity theorems still hold. So the loss of excluded middle can be restricted to where it seems to matter, to the theory of potentially infinite sequences of rational-number-surrogates.[10]

4.9.2 *Russian constructivism, and recursive analysis*

Brouwer's use of choice sequences, and the bar induction and fan theorem, are controversial. The Russian constructivists reject the very idea of a lawless sequence. To the contrary, they think of potentially infinite sequences as those generated by an *effective rule*. Invoking Church's thesis, each such effective rule is given by a recursive function, and each such recursive function can be coded with a natural number. Our second option here is to follow this lead.

Define a *Russian-real-number-generator* to be a pair of (total) recursive functions $\langle f, g\rangle$ with the following features: (i) f goes from natural numbers to (codes of) rational-number-surrogates; (ii) for each natural number n, $g(n)$ is (the code of) an interval congruent to an $(n+1)$-section of the unit interval i and left-end-equivalent to a surrogate rational number; and (iii) for any $m > n$, $g(n)$ left-overlaps $f(m)$. So, essentially, a Russian-real-number-generator is a recursive real-number-generator (in the above sense), together with a recursive function giving the relevant bounds.

[9] We mean by this that internal inconsistency would result in the Brouwerian theory of real analysis if the law of excluded middle, *written with the intuitionistic connectives*, were in play. One of us at least (Hellman, as in [55]) would insist that no conflict is posed by this with classical analysis, due to the distinct meanings of the corresponding logical connectives deployed in the different systems. See Shapiro [110], chs. 4–5, for a (possibly) different view on the meanings of the logical connectives.

[10] Thanks to David McCarty for these references, and to Mark van Atten for help with Brouwer.

We also need to modify (or refine) the above definition of equivalence accordingly. Say that the Russian-real-number-generator $\langle f_1, g_1\rangle$ is *recursively equivalent* to the Russian-real-number-generator $\langle f_2, g_2\rangle$ if there is a recursive function h such that for every natural number n, $h(n)$ is congruent to an $(n + 1)$-section of the given unit interval i and left-end equivalent to a rational-number-surrogate, and for any $m > n$, $h(n)$ left-overlaps both $f_1(m)$ and $f_2(m)$.

It is easy to check that this is an equivalence relation. A *Russian-real-number-surrogate* is an equivalence class of Russian-real-number-generators.

Define H to be a *Real-valued function* if H is a function from Russian-real-number-generators to Russian-real-number generators and for all $\langle f_1, g_1\rangle$ and $\langle f_2, g_2\rangle$, if $\langle f_1, g_1\rangle$ is recursively equivalent to $\langle f_2, g_2\rangle$, then $H(\langle f_1, g_1\rangle)$ is recursively equivalent to $H(\langle f_2, g_2\rangle)$. Intuitively, H takes real numbers to real numbers.

If the background logic is classical, then, adapting a point made just above, one can define discontinuous real-valued functions. For example, let $H(\langle f, g\rangle)$ be a surrogate for zero if f converges to a rational-number-surrogate, and let $H(\langle f, g\rangle)$ be a surrogate for one otherwise.

There are a few ways to proceed from here. Of course, one option is to accept this conclusion, and thus the existence of discontinuous functions. The theorist maintains that by identifying a real-number-generator as a recursive function, we have done all we need to in capturing the potential nature of the sequence. Each recursive function is, in fact, determined by a finite description—its code in a standard presentation of recursive functions. Moreover, the algorithm for each such function determines its entire output. And, the view goes, it is fair to apply excluded middle to this output. So we can insist, for example, that a given Russian-real-number-generator either converges to a rational number or to an irrational number.

We think it is far too much of an anachronism to even wonder if Aristotle, or an ancient or medieval or even early modern Aristotelian, would accept this perspective. Those more contemporary mathematicians who take the notion of potential infinity seriously do not accept it. The Russian constructivists themselves reject the underlying classical logic, even for the background arithmetic. They adopt an object language version of Church's thesis. Let $T(m, n, p)$ say that p is the code of a complete computation of the Turing machine with code m when it starts with input n; and if p is the code of a complete Turing machine computation, then let $U(p)$ be its output. As usual, T is a primitive recursive relation and U is a primitive recursive function on natural numbers. So both of those are expressible in the language of arithmetic. Church's thesis is the scheme:

$$(\forall m)(\exists! n)(\Phi(m, n) \rightarrow (\exists e)(\forall m)(\exists n)(T(e, m, n)\ \&\ \Phi(n, U(n)))), \qquad \text{(CT)}$$

one instance for each formula Φ in the language or arithmetic (where m and n range over natural numbers) that does not contain e free. In words, (CT) says that if, for each natural number m, there is a unique natural number n, with a certain property, then there is a recursive function that computes this m from n.

For the constructivist, a function on natural numbers *just is* a computable correspondence between numbers, where each argument has exactly one value under the computation. In a sense, then, the constructivist holds that all number-theoretic functions are computable. In this context, then, (CT) intuitively says that all (total) functions on natural numbers are (total) recursive.

It is easy to see that (CT) is inconsistent with excluded middle. Consider the following formula:

$$\alpha(m,n):\ \ ((\exists x)T(m,m,x)\ \&\ n=1)\vee(\neg(\exists x)T(m,m,x)\ \&\ n=0).$$

In words, $\alpha(m,n)$ says that either $n = 1$ and the Turing machine with code m halts when given m as input, or else $n = 0$ and the Turing machine with code m fails to halt when given m as input. With excluded middle, it follows that for each m either $(\exists x)T(m,m,x)$ or $\neg(\exists x)T(m,m,x)$—either that Turing machine, with that input, halts or it doesn't halt. So, with excluded middle, we have $(\forall m)(\exists! n)\alpha(m,n)$. So with the relevant instance of (CT), we would have that there is a recursive function whose value is 1 on input n if and only if the Turing machine coded by n halts on n. And, of course, there is no such recursive function.

The Russian constructivists also adopt a thesis now known as Markov's principle. It is a weak consequence of excluded middle. Let P be a predicate of natural numbers. The relevant instance of Markov's principle is:

$$\text{If } (\forall n)(P(n)\vee\neg P(n)) \text{ and if } \neg(\forall n)\neg P(n), \text{ then } (\exists n)P(n). \qquad \text{(MP)}$$

Markov's Principle is the thesis that if it is not possible for an algorithm to fail to terminate, then the algorithm terminates. This, of course, is classically valid, but not so in ordinary intuitionistic mathematics.

With (CT) and (MP), the Russian constructivists do prove that all functions on real numbers are continuous. The result easily ports to the present context. If we adopt intuitionistic logic, along with (CT) and (MP) in the background—even for the underlying arithmetic and the Aristotelian theory of space—then all real-valued functions from Russian real-number-surrogates to Russian real-number-surrogates are continuous (see [26], ch. 3; [120], chs. 4–6; or [12], ch. 4). But, to repeat, this requires that we invoke only intuitionistic logic, even for the original gunky Aristotelian theory and for the surrogate theories of the natural and rational numbers.

Another option, still in the spirit of this subsection, is to simply insist that real-valued functions must themselves be recursive functions (on codes for the real-valued functions). After all, we accepted (from the present perspective) that we would only countenance *recursive* sequences of rational-number-surrogates. It would seem to undermine at least the spirit of that if we then allow non-recursive functions on real-number-surrogates. That is, if we are going to insist that all real-number-surrogates are recursive sequences, it would seem that we should also insist that functions between real-number-surrogates be recursive.

It is a (classical) theorem that every real-valued function that is given by a recursive function on Russian-real-number-generators is continuous (see Rogers [100], p. 371, exercise 15-35). And, again, this is the desired result. Here we envision a perspective from which we maintain classical logic everywhere, in the background gunky theory, in the arithmetic, and even in the theory of the surrogate-real-numbers.

In sum, all but one of the theoretical options broached in this sub-section have it that all functions on the real-number-surrogates are continuous. The exception is where we assume full classical logic and take arbitrary functions from the real-number-generators to real-number-generators, a position that is, arguably, unstable. It is not clear whether this is compatible with the underlying theme of taking the Cauchy sequences to be only potentially infinite.

4.9.3 *Bishop-style constructivism*

Errett Bishop [19], and his followers, develop a version of constructivism that is a sub-theory of ordinary, classical analysis. They have no theorems that are classically false. So the Bishop follower will not have a theorem that all real-valued functions are continuous since, of course, discontinuous functions are easy to define in classical analysis. However, the constructivist also does not contradict any theorem of intuitionistic analysis or, for that matter, Russian constructivism. So they do not prove that there are any discontinuous functions. So Bishop-style constructivists are officially agnostic on the question of whether there are any discontinuous functions. And thus they do not recognize the existence of any discontinuous functions.

4.10 Continuity and Decomposability

The main theme of all aspects and instances of the present study is that none of our various gunky spaces are composed of points. Nor are they composed of (n-tuples of) real numbers. This applies, of course, to the present Aristotelian space. Its regions are not composed of real-number-surrogates, in any reasonable sense of "composed". Nor do the regions contain points, in any reasonable sense of "contained". Real-number-surrogates are equivalence classes of sequences of intervals. Like the defined "points" in the semi-Aristotelian treatment, real-number-surrogates occur well up in the type hierarchy starting with regions. Real-number-generators are superstructure, imposed on the space.

It is, of course, straightforward to define various relations between real-number-surrogates and regions. Let s be a region and let $a = \langle a_0, a_1, \ldots \rangle$ be a real-number-generator. Say that a (and the real-number-surrogate defined by a) is *interior* to s if there are intervals j and j' that are both part of s, with j' adjacent to j and left of j, and there is a natural number N, such that for any $m > N$, j left-overlaps a_m. Intuitively, a is interior to s just in case eventually, all of the left ends of the intervals in a are (well) inside s.

Say that a (and the real-number-surrogate defined by a) is *exterior* to s if there are intervals j, j', and j'' that are all discrete from s, with j' adjacent to j and left of j, j'' adjacent to j and right of j, and there is a natural number N, such that for any $m > N$, j left-overlaps a_m. Intuitively, a is exterior to s just in case, eventually, all of the left ends of the intervals in a are (well) outside s, separated from s by the intervals j' and j''.

It would be tempting to say that a is on the boundary of s just in case a is neither interior nor exterior to s. But, in the constructive context, it is better to invoke a more positive definition. So the real-number-surrogate defined by a is on a *boundary of s* if the following holds:

Let j be any interval such that there is a natural number N, such that for any $m > N$, j left-overlaps a_m. Then j overlaps s and j contains a region that is discrete from s.

Again, there is no sense in asking whether a given boundary of a given region is either in the region or not in the region. This is a straightforward category mistake.

The "space" of real-number-generators is, in at least some sense, punctiform. There is nothing "smaller than" or "inside" a given real-number-generator, and the entire "space" is composed of nothing but real-number-generators. The underlying Aristotelian space, of course, is not punctiform. That is the point (no pun intended).

As noted above, with the original "semi-Aristotelian" theory, with actual infinity in place, the imposed "points" have a structure that is remarkably similar to the underlying gunky space. In the one-dimensional case, there is a set of intervals that is isomorphic to the Dedekind–Cantor real numbers. And the topological models show that the connection goes in the other direction as well: the regular open sets of real numbers constitute a model of the gunky theory (as do the regular, closed sets of real numbers). So the punctiform Dedekind–Cantor theory is, in a sense, equivalent to the gunky theory. They are all but notational variants of each other.

The connection between the present Aristotelian one-dimensional theory and that of the imposed real-number-surrogates—the superstructure—is not as straightforward. Let us assume, for the time being, that all functions from real-number-surrogates to real-number-surrogates are continuous, as they are on most of the theoretical options broached in the previous subsections.[11] Then our Aristotelian gunky line has a very different structure than that of the "punctiform" superstructure.

If k is any interval in the space, let $O(k)$ be the set of surrogate-real-numbers that are internal to k and let $C(k)$ be the set of surrogate-real-numbers that are not exterior to k. The former is an analogue of an open interval and the latter is an analogue of a closed interval. Again, we need not broach the question of whether one of these, $O(k)$

[11] As in constructive mathematics, it follows that the axiom of choice fails. It is easy to show that for any real-number-generator a there is a real-number-generator b such that $a < b$, and b is the surrogate for a natural number. Indeed, just determine the second member of a and take any natural number that is much larger than that. It would follow from the axiom of choice that there is a real-valued function H, such that for every real-number-generator a, $H(a)$ is a rational-number-surrogate of a natural number and $a < H(a)$. But, of course, any such function is discontinuous.

or $C(k)$, is the metaphysical basis of the interval k, or if it is the other way around. It turns out that, here, the gunky interval and the sets have very different features.

Our axiom of n-sections has it that there are two discrete, congruent intervals k_1, k_2 such that k_1 is left of and adjacent to k_2, and k is the fusion of k_1 and k_2. That is, intervals in the gunky space have (clean) bisects. For any such interval, there are two congruent regions, with no part in common, whose sum is the given interval. Our ability to embed the rational numbers in the space, and, ultimately, to define the real-number-surrogates, depends on this feature of the space.

But, under the present assumptions, the real-number-surrogates do not have this feature. They exhibit a strong from of *indecomposability*. Suppose that A and B are disjoint sets of real-number-surrogates, whose union is either $O(k)$ or $C(k)$, the sets of real numbers that sort of correspond to our interval k. Then we can define a function H, on either $O(k)$ or $C(k)$, as follows: $H(a) = 0$ if $a \in A$, and $H(a) = 1$ if $a \in B$. By the prevailing assumption, H is continuous. So either $H(a) = 0$, for all a in the domain, or else $H(a) = 1$, for all a in the domain. So either A is empty or B is empty. So our "interval" cannot be divided in any non-trivial way.

If, instead, we follow the Bishop-style constructivists, we cannot show that there are no discontinuous functions. So the above argument fails. However, in line with the neutrality of the Bishop school, we also cannot show that there are discontinuous functions. So we get a weaker conclusion: one cannot show that the surrogate "open" and "closed" intervals can be divided, for that would entail the existence of a discontinuous function. Even that is enough to drive a wedge between the superstructure and the underlying gunky space, since, in the original space, intervals *can* all be divided cleanly, thanks to our n-section principle.

Aristotle himself postulated a kind of indecomposability, even for intervals, but that is due to his play with points. Once again, for Aristotle, points are the endpoints of intervals. So, according to Aristotle, when one breaks a line segment, say by bisecting it, one *creates* something new, something that only existed potentially before, namely the resulting common end–point of the two parts. In *Metaphysics* 3.5 (1002a28–b11), he writes:

> For as soon as bodies come into contact or are divided, the boundaries simultaneously become one if they touch and two if they are divided. Hence, when the bodies have been put together, one boundary does not exist, but has ceased to exist, and when they have been divided, the boundaries exist which they did not exist before (for the point, being indivisible, was not divided into two). (quoted in [116], 11)

Our space A^3, along with the original "semi-Aristotelian" space, do not share this feature. Our regions are decomposable simply because we do not recognize points at all, at least not at that level. Nothing has to be "created" or made actual, in order to bisect an interval.

Aristotle's indecomposability is thus rather different from, and much weaker than, the indecomposability that is found in the various constructive theories of the real numbers. With the latter, it is not just one "point" (or two) that undermines the decomposability of intervals. For the constructivists, there is no sense of "creating" the missing point when one tries to divide up a connected region. Instead, any attempt to bisect an "interval" of surrogate-real-numbers will *leave something out.*

Aristotle's views on points seem to be related to his thesis that continua are unities. That much, we think, is shared by the constructive theories. Regions are unified in the sense that they cannot be cleanly broken into parts.

We conclude this chapter with a brief summary. The play with infinity, whether actual or merely potential, is a crucial element in characterizing continua. In a sense, the present Aristotelian space, along with Aristotle's own space and probably Euclid's, does not recognize infinity at all. Again, there are particular *procedures* that can be iterated indefinitely, but there is no theory of infinite things of any sort. If we invoke the actual infinite, at the level of the basic, gunky space (via the axiom of fusions), then we obtain our opening, "semi-Aristotelian" space, one that is, essentially, equivalent with the now orthodox Dedekind–Cantor continuum. If, instead, we invoke a theory of potentially infinite sequences, at the level of added superstructure, we obtain an intuitionistic, constructive, or recursive account of continua, all of which are indecomposable in a rather robust sense (see Bell [13], [14], [15] for an in-depth treatment of indecomposability). It is an interesting range of options.

5

Regions-based Two-dimensional Continua: The Euclidean Case

5.1 Introduction

This chapter builds on the semi-Aristotelian theory developed in Chapter 2, recovering the so-called "classical" one-dimensional continuum on a point-free basis. Although our approach in the two-dimensional case is similar in certain respects, several new problems and issues arise demanding their own treatment. A well-known precedent for such efforts is Tarski's [119] ingenious reconstruction of three-dimensional Euclidean geometry on a point-free basis, using the primitive "sphere", in addition to first-order logic and mereology (see also [17], [36], [91]). There, however, once Tarski had defined *point* and *equidistance* of two points from a third, he simply adopted a known system of axioms for three-dimensional Euclidean geometry based on these two primitives.[1] This last step, however, is the very antithesis of "honest toil", better described as "grand larceny". Indeed, Tarski acknowledged that his system was "far from being simple and elegant".[2] Moreover, he wrote that "it seems very likely that this postulate system can be essentially simplified by using intrinsic properties of the geometry of solids".

[1] The system is due to Pieri [93].

[2] Tarski here severely understates the problem with his procedure. The usual Nagelian (e.g. [88], [89]) standard for reducing a theory T_2 to another T_1 requires both (1) that the predicates of T_2 be explicitly defined by those of T_1, *and* (2) that the axioms of T_2 be derived as theorems of T_1 using the definitions. Here (2) serves as a substantive constraint on the definitions of (1). But Tarski did not provide axioms for his T_1, beyond those of pure logic and mereology, and some postulates on "solid", defined as the fusion of a nonempty set of "spheres". Tarski simply adopted the translates of axioms of T_2 as new axioms of T_1. By such a procedure, any theory is automatically reducible to any other (with enough predicates of the right arity), for nothing constrains the "definitions"!

Perhaps this has been missed because, in fact, Tarski's definitions of external and internal tangency and diametricality, concentricity, etc. aren't at all arbitrary but rather qualify as good and ingenious, despite his failure to state regions-based axioms on 'sphere'. Our evidence for this is that, within the point-based theory (T_2), we can derive theorems to the effect that each of the predicates to be defined holds if and only if the proposed *definiens* holds, but where the latter is translated back into the point-based theory. (No doubt, such reasoning is what led Tarski to contrive his definitions in the first place.) However, that reasoning all takes place in the "superstructure", that is in T_2, and so provides no substitute for the Nagelian requirement (2).

See [49] for an in-depth treatment of Tarski's definitions and axioms.

Here we will present an "honest toil" regions-based reconstruction of two-dimensional continua and Euclidean geometry. To aid the exposition, we will proceed in stages, first deriving the Archimedean property for a class of readily postulated orientations of certain special regions, "generalized quadrilaterals" (intended as parallelograms), by which we cover the entire space. The second stage consists in generalizing this to arbitrary orientations, and then establishing an isomorphism between the space and the usual point-based $\mathbb{R} \times \mathbb{R}$.

As in the one-dimensional case, this is done on the basis of axioms which contain no explicit "extremal clause" (to the effect that "these are the only ways of generating regions"), and we have no axiom of induction other than ordinary numerical (mathematical) induction. Finally, having explicitly defined '*point*' and '*line*', we will derive the characteristic Parallel's Postulate (Playfair axiom) from regions-based axioms, and point the way toward deriving key Euclidean metrical properties. The chapter closes with a brief sketch of how to extend the treatment to three dimensions, and beyond. No new conceptual issues arise.

5.2 Regions-based Two-dimensional Continuum: Derivation of the Archimedean property (Restricted)

As in the one-dimensional case, we begin with first-order logic with =, augmented with plural quantification (or second-order logic) and mereology, taking over axioms 1a, 1b, and 2 from the semi-Aristotelian treatment:[3]

1a. Axioms on $x \leq y$ ("x is part of y"): reflexive, antisymmetric, transitive.

1b. Axiom on $\leq$ **and** $\circ$: $x \leq y \leftrightarrow \forall z[z \circ x \rightarrow z \circ y]$.

Fusion or whole comprehension:

$$\exists u\Phi(u) \rightarrow [\exists x\forall y\{y \circ x \leftrightarrow \exists z(\Phi(z) \,\&\, z \circ y)\}],$$

where Φ is a predicate of the second-order language (or language of plurals) lacking free x.

The parts of our space, which constitute the range of our first-order variables, we again call "regions". Informally, we'll sometimes refer to our space as G^2. Nominalistic summing or fusion is indicated by '+', which applies to arbitrary regions, with $r+s = t \iff \forall u[u \circ t \leftrightarrow (u \circ r \vee u \circ s)]$. As before, sums or fusions of all things of any

[3] Recall that we follow standard practice in defining a relation $x \circ y$, called "overlaps":

$$x \circ y \Leftrightarrow^{df} \exists z(z \leq x \,\&\, z \leq y).$$

We write $x|y$ for $\neg\exists z[z \leq x \,\&\, z \leq y]$, pronounced "$x$ is discrete from y". Axioms 1a and 1b imply the Extensionality Principle:

$$x = y \leftrightarrow \forall z[z \circ x \leftrightarrow z \circ y].$$

(non-empty) plurality (or satisfying any non-vacuous predicate) are recognized. We use $\sum_{n=0}^{\infty} x_n$ to designate fusions of infinitely many regions. In the case of predicates or formulas, quantification over arbitrary wholes or pluralities is permitted, so that the system is in this sense "impredicative", as expected in a classical reconstruction (see the final section of Chapter 3).

At this stage, our only other primitives are as follows: First, we have a primitive for *congruence* of regions, $Cong(s, r)$, understood as it usually is, to mean "same shape and size". We also have a special type of region which we call "generalized quadrilateral" $gq(r)$. Our axioms will in effect delimit these to parallelograms, although we don't use that term, as the notion of "line" is not yet available.[4]

We also introduce four relative "direction" primitives associated with any given gq viz. 'Up', 'Down', 'Left', and 'Right' (using uppercase first letters as abbreviations), subject to axioms to be introduced below. We also define a notion of "co-oriented adjacency" of gq's, $Adj(r, s)$ iff $r|s$ & $gq(r + s)$. Intuitively, this means that r and s share exactly one entire border in common; that either s is U of r and r is D of s, or vice versa, or r is L of s and s is R of r, or vice versa; that there are no gaps between r and s; and that they are oriented in the same way, not oppositely, if they are not bilaterally symmetric. (In the latter case, if they are not both rectangles, their sum would not form a gq, despite sharing an entire border in common.) We normally drop the reference to co-orientation.

Our first new axiom specifies obvious properties of congruence and the second introduces generalized quadrilaterals:

Axiom 2.0: $Cong(s, r)$ is an equivalence relation.

Axiom 2.1: There exist gq regions. Any gq region, r, has a "*quadra-section*", that is, $r = s_1 + s_2 + s_3 + s_4$, each itself a gq, with all the s_i discrete from one another, with $Cong(s_i, s_j)$, and such that for each s_i, there are exactly two s_j such that $Adj(s_i, s_j)$.

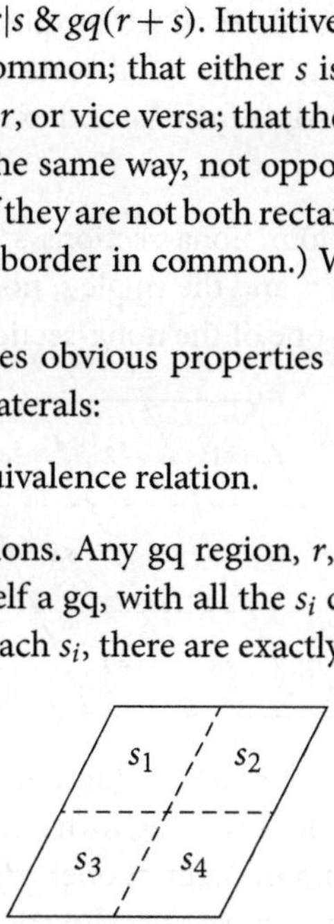

[4] In the next section we generalize the Archimedean property to a plenum of directions and derive the Euclidean parallels postulate (EPP). There, we will introduce further primitives pertaining to *angles* (or to angles less than 2π radians), conceived informally as regions shaped as sectors of circles (although we do not need 'circle' as a primitive).

Notice, incidentally, that in the one-dimensional theory of the previous section, "congruence" is only defined for intervals; here it is a relation on all regions. In the present section, however, we only apply "congruence" to generalized quadrilaterals. In the next section, "congruence" is also applied to "angles", construed as sectors of circles, and to "triangles", which are certain parts of gq's. For other regions, we can informally think of "congruence" as identity, or, for that matter, any other equivalence relation.

Furthermore, r has a "*quadra-extension*", that is, there are *exactly* four gq's, r_i, pairwise discrete and discrete from r, with $Cong(r_i, r)$ and $Adj(r_i, r)$ for $i = 1, \ldots, 4$, such that each r_j is adjacent to the fusion of exactly two s_i, and each s_i enters into exactly two such fusions.

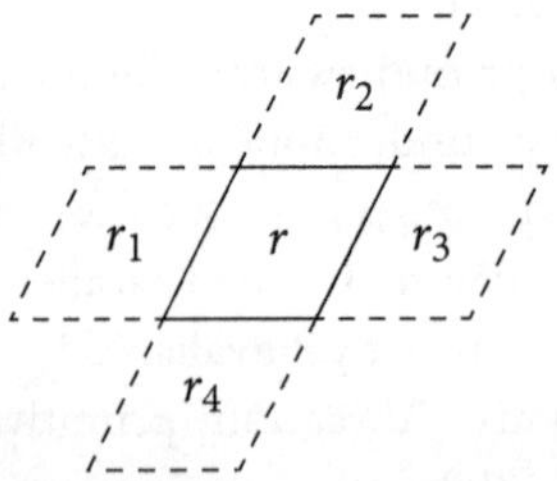

Finally, r has a "*nona-section*", that is, $r = \sum_{i=1}^{9} s_i$, each s_i a gq, with $s_i|s_j$, $i \neq j$, and $Cong(s_i, s_j)$. Each extension r_i of r determined by quadra-extension is adjacent to the fusion of exactly three sections s_i of the nona-section; each of exactly four of the s_i enters into exactly two of these triple fusions, and exactly one of the s_i enters into no such triple fusion.

To aid visualization: The first four nona-sections s_i of the final clause, informally speaking, are "at the corners" of r, and the single s_i not entering into any triple fusion adjacent to an r_j is the "central" one of the nona-section.

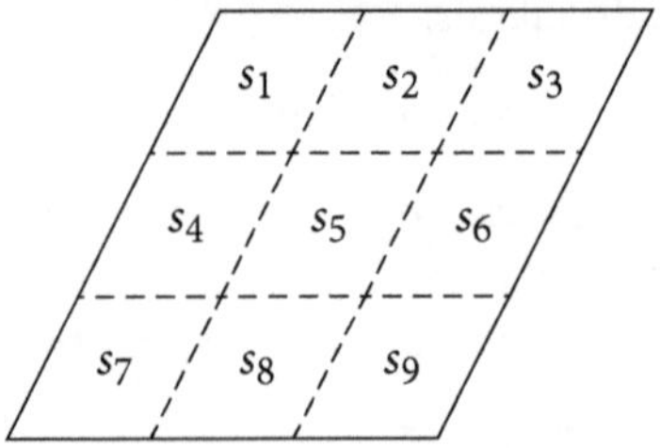

Then repeated quadra-extension results in (among other things) four sums (call them "strings") of congruent regions starting with a central one, each string arranged linearly with each region (beyond the central one) adjacent to its predecessor and its successor, proceeding outward in a particular direction (the "direction of the string") corresponding to those of the principal axes of the central gq-region. A *segment* of such a string—equal to the fusion of the initial region and some of its successors and closed under "immediate predecessor"—is called "*bounded*" just in case there is some region discrete from all the members of the segment and beyond all of them in the same direction as that of the string. Our next axiom stipulates that sectioning and extending are inherited by congruents and certain sums thereof, and that these "operations" are indefinitely iterable:[5]

[5] Strictly speaking, we are specifying what regions exist, regardless of the application of operations. But speaking of operations is natural and perhaps of heuristic value.

Axiom 2.2: Any region congruent to a gq is itself a gq.

Below we will introduce "rectangles" as special kinds of gq's and will adopt as an axiom (Axiom 2.12e) that every region has a rectangle as a part. Thus our axioms guarantee infinite divisibility of the space into gq-regions. Furthermore, they guarantee that the space is infinite in extent in all directions, and that we can find gq-regions (intuitively) at arbitrary finite distances from any given one. Of course, we do not yet have that the space is Archimedean.

Note that Axiom 2.1 embodies the closure of gq's under fusion of a gq and any one of its extensions, by the requirement of adjacency as defined. One further closure condition on gq-regions will prove useful in recovering the classical Dedekind–Cantor continuum as superstructure over any "line" in our space, once that notion has been defined. To state this, define a sequence, $\rho = \langle r_j \rangle$, of gq's to be "*sequentially adjacent*" just in case, for each j, $Adj(r_{j+1}, r_j)$, and for some direction D, $r/j + 1$ is D of r/j, for all j. (More axioms governing "directions" are given below.) Also, denote by $\Sigma_j(\langle r_j \rangle)$ the fusion of the gq's, r_j, of such a sequence. Then we stipulate:

Axiom 2.3a: If $\rho = \langle r_j \rangle$ is sequentially adjacent and for some gq r, and every j, $r_j \leq r$, then $\Sigma_j(\langle r_j \rangle)$ is a gq.

This is, in effect, a Cauchy completeness condition adapted to gq-regions, and it will guarantee that, once the relevant terms have been suitably defined, gq's with "borders" of arbitrary real "length" always exist.

Because of the iterability of sectioning, we are also furnished with many nested sequences of gq-regions, and many of these are convergent (in standard terms, to a point).

gq-regions $\langle r_i \rangle$ are *nested* iff $r_k < r_j$ for $k > j$.

For point-wise convergence, however, we need a stronger nesting property:

gq-regions $\langle r_i \rangle$ are *properly nested* iff r_{k+1} is one of the sections of quadra-secting or nona-secting r_k, for all $k = 1, 2, \ldots$

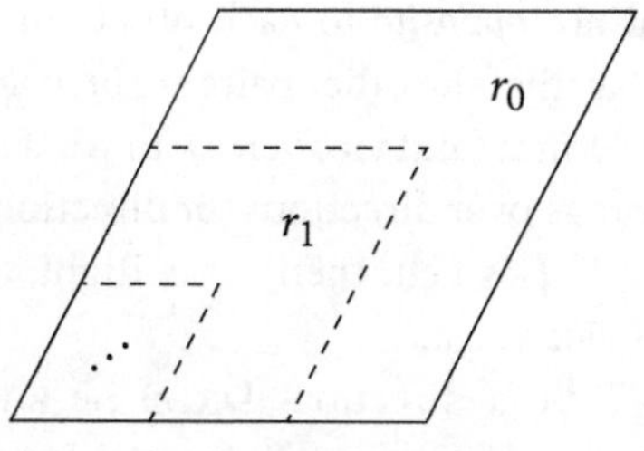

We will also, for convenience, sometimes call properly nested sequences "convergent", although we cannot yet distinguish true "convergence to a point" from "convergence to an infinitesimal region", not yet having proved that our space is Archimedean.

Theorem 2.1. *There are properly nested gq-regions.*

Proof. By Axiom 2.1, we may, for instance, take r_1 of our sequence to be, say, the UL, upper left quadrant of a first quadrasection of a given gq-region, r. Given r_i, then let r_{i+1} be the DR, down-right quadrant of r_i. This defines a properly nested sequence (intuitively converging to the DR-most corner of the NW quadrant of r, i.e. the center of r). (Alternatively, work with repeated nona-sections to obtain a properly nested sequence converging in the same way, by always choosing the central region of the nona-section at any stage.) ■

Let r be a gq-region. We now introduce an operation, based on quadra-extension, that provides larger and larger regions, starting with r, growing exponentially without finite bound. Congruents of parts of any of these will then be "finite relative to r". The operation, called "4-fold-extension", is best illustrated with an example: Start with r, with its own relative orientation, labeled 'U', 'R', 'D', 'L'. Quadra-extend r and designate by r_U the congruent of r adjacent to r and up, and similarly for r_R. Now by r_{R_U} designate the extension of r_R congruent and adjacent to it and up from it (selected to agree with "up" from r). Then the *4-fold-extension of* r—"4-$f(r)$"—is the gq-region $r + r_R + r_{R_U} + r_U$. Intuitively, the area of 4-f(r) is four times that of r. By iterating 4-f we specify exponentially larger and larger gq's, intuitively occupying the UR quadrant of our space. (If the space is Archimedean, all of that quadrant will be covered by these regions.)

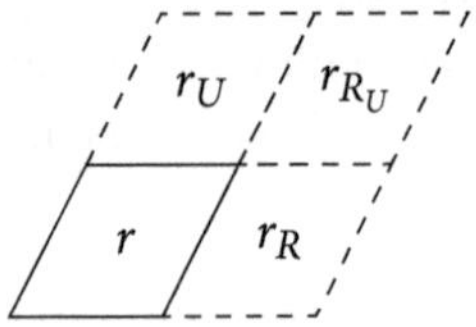

As indicated above by the choice of names of "directions", these terms are not absolute but are relative to each gq. In particular, the various gq's need not be "oriented" the same way (although we take some steps to orient some of them, for purposes of simple exposition).

Say that Left and Right are *opposite* to each other, and also that Up and Down are *opposite* to each other. Say that the other pairs are *orthogonal.* So Up and Down are each orthogonal to Left and Right (and vice versa). In what follows, we'll use the meta-variables T, T', T_1, etc. to range over directions (or direction labels). If T is a direction, then $T-$ is its opposite. So if T is Left, then $T-$ is Right, and vice versa, and if T is Up, then $T-$ is Down, and vice versa.

Let s be a gq and let T be a direction. Let m be any region. We want some axioms governing the relation of m being T from s. Intuitively, the idea is that m is discrete from s and lies entirely on the T side of s, between lines that form the opposite borders of s (not necessarily strictly between). Of course, we cannot say that officially, since there are no lines (and so no borders). Instead, we give some axioms governing the relation. Of course, in proving the theorems, we rely only on the axioms.

Officially, we have four new binary relations between gq's and regions: *Left*(s, m), *Right*(s, m), *Up*(s, m), and *Down*(s, m). But we will usually speak more informally of a given region m being T of a gq s, where T is one of the four directions.

Axiom 2.4a: Let m be a region, s a gq, and T a direction. If m is T from s, then $m|s$.

It follows immediately that the relations *m is T of s* are irreflexive.

Axiom 2.4b: Let m, m' be regions, s a gq, and T a direction. If m is T of s and $m' \leq m$, then m' is T of s.

Axiom 2.4c: Let s be a gq, T a direction, and let M be any set (or plurality) of regions. If every member of M is T of s, then the fusion of M is also T of s.

Suppose that m is T from s. Then Axiom 2.4a says that the two are discrete and Axiom 2.4b says that every part of m is also T from s. Axiom 2.4c is a sort of converse to Axiom 2.4b.

Next are axioms that relate the four directions to the quadra-extensions, quadra-sections, and nona-sections of a given gq. Here we coordinate some of the directions.

Axiom 2.5a: If s is a gq and T is a direction, then exactly one of the four quadra-extensions of s is T of s. So we can speak of the T quadra-extension of s.

Axiom 2.5b: If t is the T quadra-extension of s, then s is the $T-$ quadra-extension of t. Moreover, the T quadra-extension of t is also T of s (and thus the $T-$ quadra-extension of s is $T-$ of t).

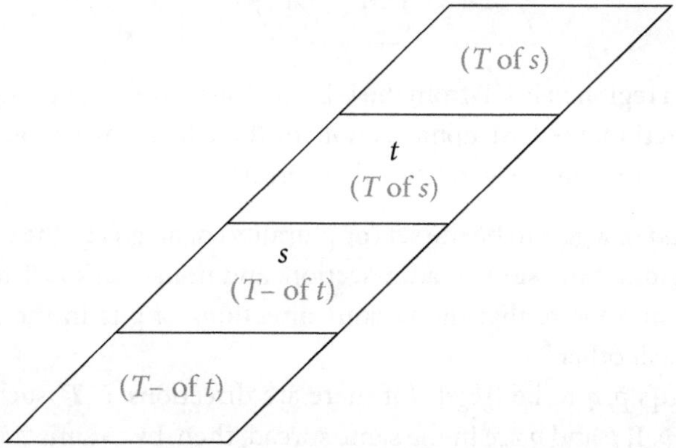

Axiom 2.6a: Let s be a gq. Then the four quadra-sections of s can be labeled Up-Left (UL), Up-Right (UR), Down-Left (DL), and Down-Right (DR), such that (i) the UL-quadra-section of s is Left from the UR-quadra-section of s and Up from the DL-quadra-section of s; the UR-quadra-section of s is Right from the UL-quadra-section of s and Up from the DR-quadra-section; and similarly for the other two.

And (ii) the Left quadra-extension of the UL-quadra-section of *s* is the UR-quadra-section of the Left quadra-extension of *s*; and the Up quadra-extension of the UL-quadra-section of *s* is the DL-quadra-section of the Up quadra-extension of s; and similarly for the other three quadra-sections of *s*.

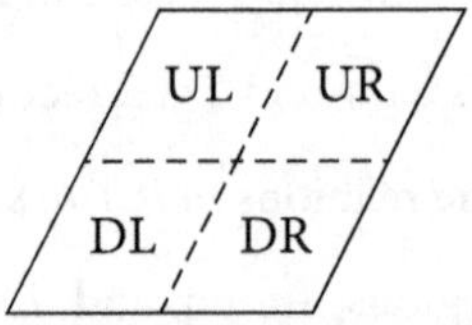

Axiom 2.6b: Let *s* be a gq. Then the nine nona-sections of *s* can be labeled Up-Left (UL), Up-Middle (UM), Up-Right (UR), Middle-Left (ML), Middle-Middle (MM), Middle-Right (MR), Down-Left (DL), Down-Middle (DM), and Down-Right (DR), such that (i) the Right quadra-extension of the UL-nona-section of *s* is the UM-quadra-section of *s*, (ii) the Left quadra-extension of the UL-nona-section of *s* is the UR-nona-section of the Left quadra-extension of *s*, (iii) the Up quadra-extension of the UL-nona-section of *s* is the DL-nona-section of the Up quadra-extension of s, (iv) and the Down quadra-extension of the UL-nona-section of *s* is the ML-nona-section of *s*; and similarly for the other eight nona-sections.

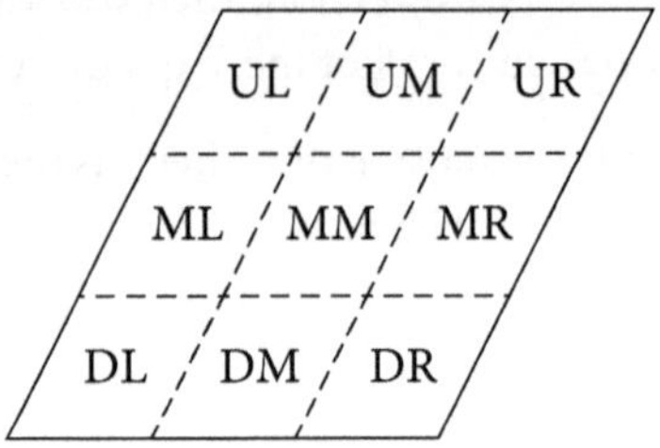

Axiom 2.6c: If a region *m* is UP from the UL-quadra-section, the UR-quadra-section, the UL-nona-section, the UM-nona-section, or the UR-nona-section of a gq *r*, then *m* is UP from *r*. And similarly for the other directions.

Define the *spread* of a gq *r* to be the set (or plurality) of all gq's in the minimal closure of $\{r\}$ under quadra-extension, quadra-section, and nona-section. The above axioms (are meant to) make sure that the various directions of gq's in the spread of *r* are oriented with each other.[6]

Define two gq's p, q to be *Aligned* if there are directions T, T' such that p is T of q and q is T' of p. If p and q are in the same spread, then, by axioms 2.5, T' will be T–, the opposite direction of T, but that need not hold in general.

[6] We could extend the coordination a bit. Let q be a gq and let r be its Right quadra-extension. Then $q + r$ is itself a gq. We can stipulate that the UL-quadra-section of $q + r$ is the fusion of the UL-quadra-section and the UR-quadra-section of q. And so on. But there is no hope of coordinating *all* of the gq's in G^2, for some of them might not even have the same "orientation" as q. Intuitively, some gq's might be at an angle to others—so to speak. So we will have to live with the relativity of the directions.

For example, let p be a gq and let q be the Up quadra-extension of the Up quadra-extension of p. Then p and q are Aligned.

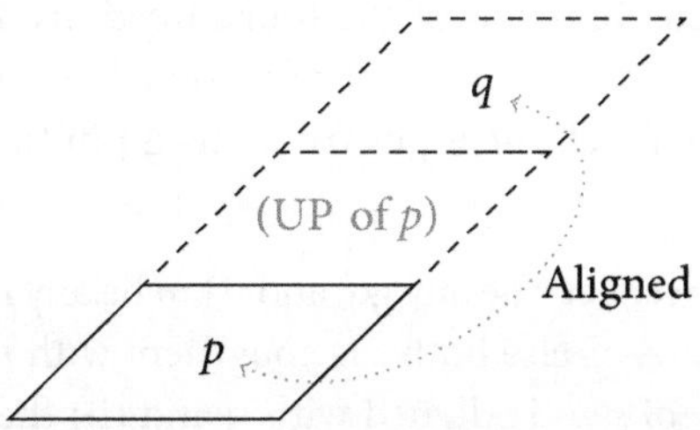

Recall that, intuitively, a region m is, say, Left of a gq s just in case m is entirely to the Left of s, and fits between the extensions of the upper and lower boundaries of s. So, intuitively, two gq's are Aligned if they are discrete and the lines formed by two of their opposite boundaries are the same. Of course, this is all at the level of heuristics.

Axiom 2.7a. Transitivity: Let p be Aligned with q. So there are directions T, T' such that q is T of p, and p is T' of q. Then any region that is T' - of q is also T of p.

In other words, if p is Aligned with q, then, in effect, the T'– direction from q is the same as the T direction from p; anything that is in the T'– direction from q is also T of p. This also guarantees that our original gq p is not T'– of q. In effect, the Transitivity axiom guarantees that the space is not "closed" or, in other words, no matter how far one goes in the T direction, one will not get back to the gq at which we started. (As already noted, Axiom 2.4a implies that 'T of' is irreflexive.)

Axiom 2.7b. Dichotomy: Let p be Aligned with q. So there are directions T, T' such that q is T of p; and p is T' of q. Let m be any region that is T' of q and discrete from p. Then either m is T of p, or m is T– of p.

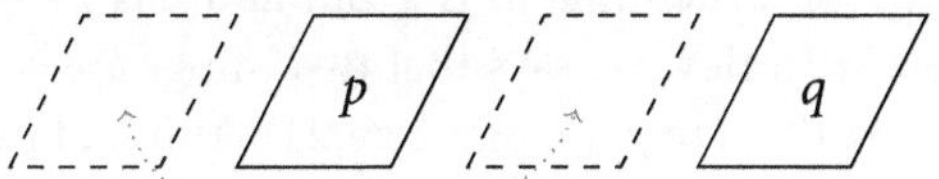

two (among other)
possible locations of m

In words, the Dichotomy axiom says if p is Aligned with q, so that there are directions T, T' such that q is T of p, and p is T' of q, then any region that is T' from q and discrete from p is either in between the gq's or on the opposite side of p.

Axiom 2.8, gq covering: Let T be a direction, and suppose that a region m is T from a gq r. Then m is a part of the fusion of all gq's that are T of r, congruent with r, and Aligned with r.

In other words, we can cover any region that is T from r with gq's, all of which are Aligned with r, congruent with r, and T of r. Notice that a converse of Axiom 2.8 follows from Axiom 2.4.

We need to rule out models in which the universe consists of two completely discrete "spaces", where no part of one of them can be "reached" from any part of the other via some combination of the four directions. For example, we should rule out models that consist of two copies of a point-free space, where no region in one of them is Up, Down, Left, or Right from any gq in the other. Our next axiom accomplishes this:

Axiom 2.9, Unity of Space: Let r be any gq, and let m be any region. Then there is a gq s such that (i) either $s = r$, or else both s is congruent with r and there is a direction T_1 of s such that r is T_1 of s and Aligned with s; and (ii) there is a direction T_2 of s, such that m overlaps the fusion of s and all gq's that are T_2 of s.

Intuitively, the idea is this: Start with any gq r and let m be any region whatsoever. Then we can reach part of m by going in some direction from r, turning either right or left (or up or down), and then going in that direction.

Notice that nothing so far entails that the space is Archimedean. Even with Axiom 2.9, we have not ruled out models that have gq's r, s where, say, s is Left of r, but infinitely far from r (relative to r, and relative to s).[7] We have also not ruled out gq's that are infinitely large relative to another. Ruling out such models, via an Archimedean theorem, is the main order of business of this section.

More definitions: Let r and s be gq's that are Aligned with each other, and let m be any region. Say that m is *Between* r and s if and only if both of the following hold:

(i) there is a direction T_1 of r such that m and s are both T_1 from r; and
(ii) there is a direction T_2 of s such that m and r are both T_2 of s.

It follows that if m is Between r and s, then r is Aligned with s.

Let m and n be regions. Say that m and n are *Contiguous* just in case there is no gq t and no direction T of t, such that m is T of t and n is $T-$ of t. This definition is loosely modeled after Aristotle's. He says that two things are contiguous just in case nothing can get between them (e.g. *Physics* 226b21). Think of two adjacent books on a tightly packed shelf or two adjacent houses whose outer walls touch (or overlap). Here, we'd *like* to say that two regions are contiguous if there can be no region that comes between them. But we need at least one gq in order to have an orientation or a "direction", and thus a sense of "between". So we say that two regions are Contiguous if no gq can separate them, so that one of the regions is on one side of the gq, and the other region is on the other side.

As discussed in Chapter 1, Aristotle defines two things to be *continuous* if, when they are contiguous, the boundary between them is absorbed, and they become one

[7] To construct such a model, begin with a two-dimensional non-Archimedean space, such as one constructed from Robinson-style hyper-reals. Let the "regions" be regular open sets (those that are identical to the interior of their closure), and let the gq's be the interiors of parallelograms.

thing.[8] Think of bodies of water or, for that matter, regions of space. If they are contiguous with each other, they are (or become) one.

We can get something similar here, at least for gq's. Let r and s be gq's. Say that r is *Continuous* with s just in case they are Aligned, Contiguous, and there is no region that is between them.

Recall that we have another primitive Adj(s, r), whose intended meaning is that r and s are discrete from each other and share part of a border, but, again, that is at the level of heuristics. It is part of the axiom defining gq's that quadra-extensions are adjacent to one another. We can relate that to the present primitives:

Axiom 2.10: Every gq is Continuous with its quadra-extensions. In general, if r and s are Aligned gq's such that Adj(r, s), then r and s are Continuous.

Notice, however, that it can happen that two gq's are congruent with each other, Aligned, and *Contiguous*, and still there is a region that is between them—so that they are not Continuous. Think of two congruent parallelograms side by side, but oriented in opposite directions. We define rectangles to be gq's where this does not happen:

Let q be a gq. Then q is a *Rectangle* just in case for every gq r, if r is Aligned with q, Contiguous with q, and congruent with q, then r is Continuous with q (i.e. there is no region that is between r and q). In fact, r is a quadra-extension of q.

Axiom 2.11a: If a region q is congruent with a rectangle q', then q is a rectangle.

Axiom 2.11b: If r and r' are Aligned, Contiguous rectangles, then $r + r'$ is a rectangle.

Some properties of rectangles:

Axiom 2.12a Decomposition I: Let r, s, t be rectangles such that r is Aligned with both s and t and that $s \circ t$. Then either $s \leq t$, $t \leq s$, or there are three pairwise discrete rectangles s', t', v, all Aligned with r such that $s = s' + v$ and $t = t' + v$. Moreover, v is between s' and t'.

For example:

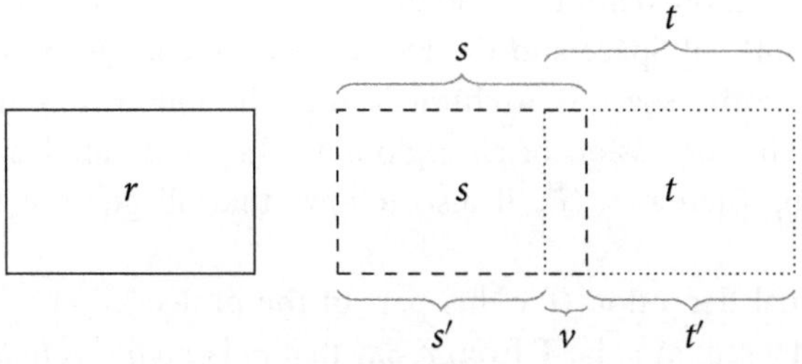

[8] *Physics*, Book V (227a6): "The continuous is just what is contiguous, but I say that a thing is continuous when the extremities of each at which they are in contact become one and the same and are (as the name implies) contained in each other. Continuity is impossible if these extremities are two. This definition makes it plain that continuity belongs to things that naturally, in virtue of their mutual contact form a unity."

Axiom 2.12b, Decomposition II: Let q and r be any gq's (or we can restrict this to rectangles). If $q < r$ then $\neg$ Cong(q, r).

Axiom 2.12c, Decomposition III: If q is a rectangle, then so are its quadra-sections and nona-sections.

Axiom 2.12d, Additivity: Let r, r_1, r_2, s, s_1, and s_2 be rectangles, such that $r = r_1 + r_2$ with r_1 discrete from r_2, and $s = s_1 + s_2$ with s_1 discrete from s_2. If Cong(r_1, s_1) and Cong(r_2, s_2) then Cong(r, s).

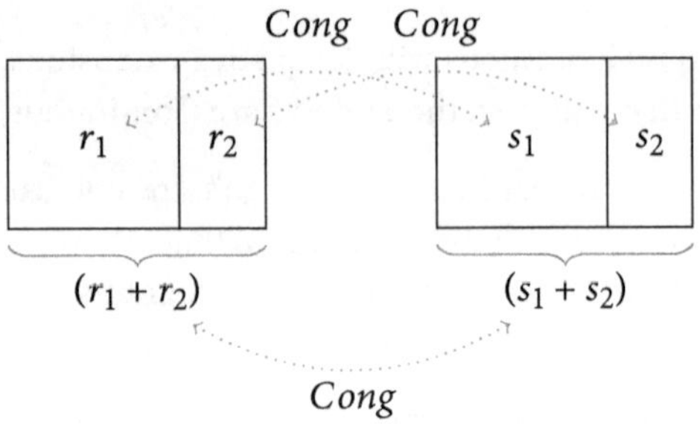

Our next axiom guarantees an abundance of rectangles, including arbitrarily small ones (in light of Axiom 2.1), as mentioned above:

Axiom 2.12e, Existence of Rectangles: Let m be any region. Then there is a rectangle r such that $r < m$.

This axiom also guarantees that the space is gunky.

Here, then, is our plan for establishing the Archimedean property: Let r be any rectangle and let T be a direction from r. Let r^T be the fusion of all regions that are T of r. Let r_1 be the T quadra-extension of r, and let R_T be the minimal closure of $\{r_1\}$ under the operation of taking the T quadra-extension. Notice that every rectangle in R_T is in the spread of r, and so has the same "orientation", by Axiom 2.6a. Let r_T be the fusion of R_T.

After adding one more definition and one more axiom, we will show that $r_T = r^T$. Then, with the Unity of Space and the Existence of Rectangles axioms, we are able to show that the entire space is Archimedean in the following sense: Let r be any rectangle. Let r^* be the fusion of all regions in the minimal closure of $\{r\}$ under quadra-extension. Then $r^* = G^2$. It also follows that all gq's are finite relative to each other.

Here is our final definition (for this part of the project): Let r be a gq and T a direction, and let a region m be T from r. Say that m is *Bounded to the* T of r if there is a gq s, such that (i) s is T of r and Aligned with r, and (ii) there is a direction T' of s such that r and m are both T' of s. In other words, m is Bounded to the T of r just in case there is a gq s Aligned with r, such that m is Between r and s.

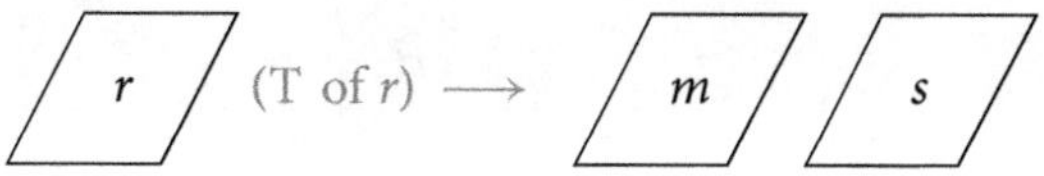

And the final axiom of this section is this:

Axiom 2.13, Translation: Let r be a gq and T a direction. Suppose that a region m is T of r. Then, if m is Bounded to the T of r, then there is a gq r' such that (i) r' is congruent to r and Aligned with r; (ii) m is between r and r'; and (iii) m and r' are Contiguous.

Intuitively, the antecedent of the Translation axiom is that m is T from r and there is some gq on the "other side" of m (from r). The consequent of the axiom is that we can get a copy of r on the "other side" of m, Aligned with r and move it so close to m that no gq can get between them. In a sense, we have this copy of r just touching m.

Theorem 2.2. *Let r be any rectangle and let T be any direction from r. Let r^T be the fusion of all regions that are T of r. Let r_1 be the T quadra-extension of r, and let R_T be the minimal closure of $\{r_1\}$ under the operation of taking the T quadra-extension. Let r_T be the fusion of R_T. Then $r_T = r^T$.*

Diagram for r^T:

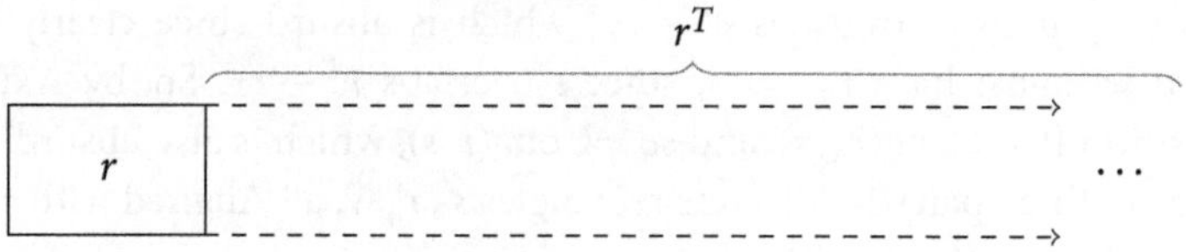

Diagram for r_T:

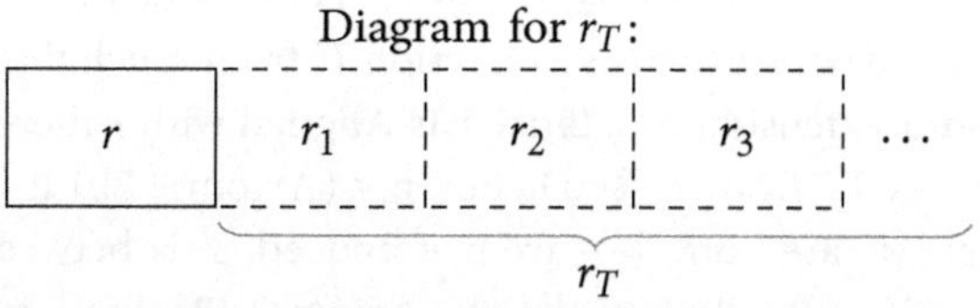

Proof. By transitivity, every member of R_T is T of r, and so, by Axiom 2.4c, r_T is T of r. So $r_T \leq r^T$. Suppose $r_T \neq r^T$. Then $r^T - r_T$ exists. By Axiom 2.4b, $r^T - r_T$ is T of r.

Let S be the set (or plurality) of all gq's that are T of r, congruent with r, and Aligned with r. The members of S are all rectangles. By Axiom 2.8 (gq covering), $r^T - r_T$ is part of the fusion of S. So let s be any rectangle in S that overlaps $r^T - r_T$. Since s is Aligned with r, there is a direction T' from s such that r is T' of s.

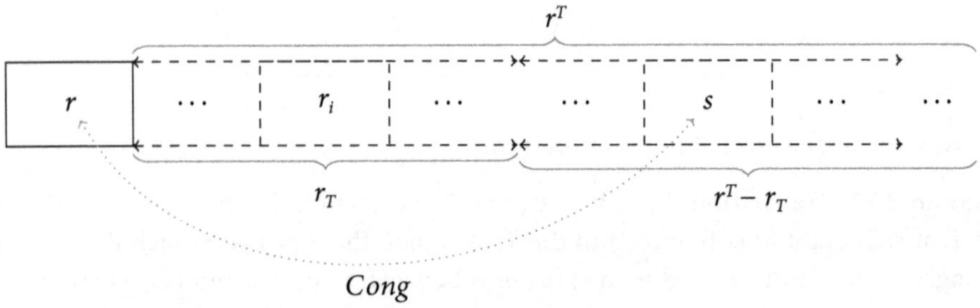

There are two cases. Case 1: Assume that s overlaps r_T. Then there is a rectangle r_n in R_T such that s overlaps r_n. So we have that s and r_n are both Aligned with r, and all three are rectangles (invoking Axiom 2.11a for r_n). So by Axiom 2.12a, either $s \leq r_n$, $r_n \leq s$, or there are three pairwise discrete rectangles s', r'_n, v, all Aligned with r, such that $s = s' + v$ and $r_n = r'_n + v$, with v between s' and r'_n.

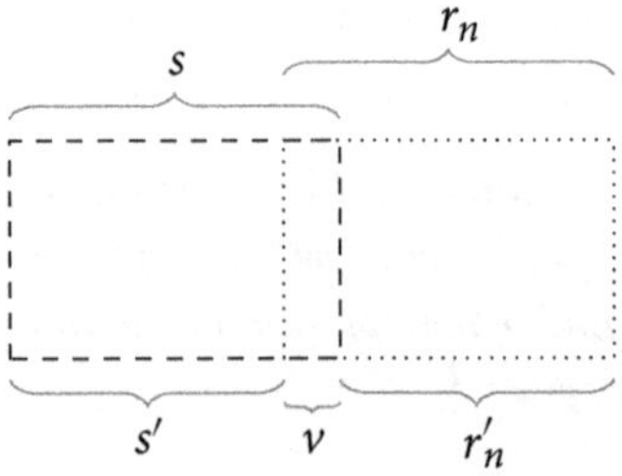

But if $s \leq r_n$, then r_n overlaps $r^T - r_T$, which is absurd since clearly $r_n \leq r_T$. If $r_n \leq s$, then we must have $r_n < s$, since s overlaps $r^T - r_T$. So, by Axiom 2.12b, Decomposition II, $\neg$Cong(r_n, s), and so $\neg$Cong(r, s), which is also absurd.

So there are three pairwise discrete rectangles s', r'_n, v, all Aligned with r, such that $s = s' + v$ and $r_n = r'_n + v$, with v between s' and r'_n.

Since $v \leq r_T$, we have that s' overlaps $r^T - r_T$. Let r_{n+1} be the T quadra-extension of r_n. Since s is Aligned with r, there is a direction T' from s such that r is T' from s. Let s'' be the T'– quadra-extension of s. Since v is Aligned with r there is a direction T'' from v such that r is T'' from v. By Dichotomy (Axiom 2.7b) it is straightforward to show that s' and s'' are both T''– from v. Indeed, s' is between v and s''. So s' is Bounded to the T''– of v. By Translation (Axiom 2.13), there is a gq v' such that (i) v' is congruent to v and Aligned with v (and so with r, and everything else here); (ii) s' is between v and v'; and (iii) s' and v' are Contiguous. By Axiom 2.11b, s' and v' are Continuous with each other and $s' + v'$ is a rectangle. By additivity (Axiom 2.12d), $s' + v'$ is congruent with s (which, recall, is $s' + v$). So $s' + v'$ is congruent with r and thus with r_n. It is also a straightforward consequence of Dichotomy (Axiom 2.7b) that $s' + v'$ is adjacent to r_n and T from r_n. So, by the uniqueness of T quadra-extension, we have that $s' + v'$ is r_n. But then $s' \leq r_T$, which contradicts the claim that s' overlaps $r^T - r_T$.

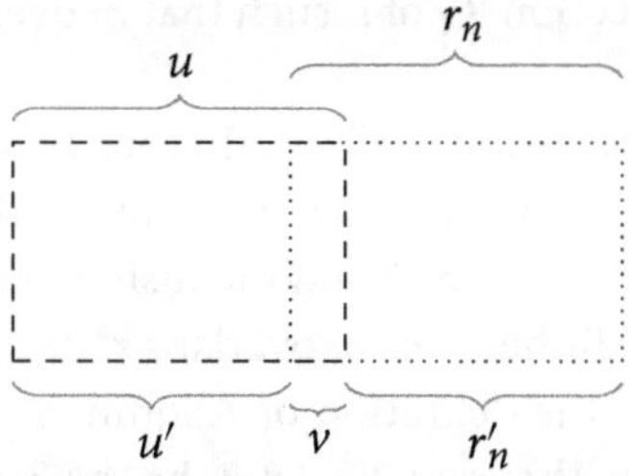

Case 2: Assume that s does not overlap r_T. By Axiom 2.7b (Dichotomy) either r_T is T' of s or r_T is $T'-$ of s. But the latter is absurd. We would have s coming between r and its T quadra-extension r_1. So r_T is T' of s. So r_T is Bounded to the T of r. So, by Axiom 2.13 (Translation), there is a rectangle t such that (i) t is congruent to r and Aligned with r; (ii) r_T is between r and t; and (iii) r_T and t are Contiguous. So there is a direction T'' such that r is T'' of t. Let u be the T'' quadra-extension of t. So u overlaps r_T (by the contiguity of t and r_T). So there is a rectangle r_n in R_T such that u overlaps r_n. If either $r_n \leq u$ or $u \leq r_n$ then $r_n = u$ (by Axiom 2.12b, Decomposition II). But in that case, by the uniqueness of quadra-extension, t would be the T quadra-extension of r_n and so $t \leq r_T$, which contradicts the hypothesis (ii) that r_T is between r and t.

So there are three pairwise discrete rectangles u', r'_n, v, all Aligned with r, such that $u = u' + v$ and $r_n = r'_n + v$, and v is between u' and r'_n. Proceeding similarly to Case 1, note that since v is Aligned with r there is a direction T_1 from v such that r is T_1 from v. By Dichotomy (Axiom 2.7b) it is straightforward to show that u' and t are both T_1- from v. Indeed, u' is between v and t. So u is bounded to the T_1- of v. By Translation (Axiom 2.13), there is a gq v' such that (i) v' is congruent to v and Aligned with v (and so with r, and everything else here); (ii) u' is between v and v'; and (iii) u' and v' are Contiguous. By Axiom 2.11b, u' and v' are *Continuous* with each other and $u' + v'$ is a rectangle. By additivity (Axiom 2.12d), $s' + v'$ is congruent with u (which, recall, is $u' + v$). So $u' + v'$ is congruent with r and thus with r_n. It is also a straightforward consequence of Dichotomy (Axiom 2.7b) that $u' + v'$ is adjacent to r_n and T from r_n. So, by the uniqueness of T quadra-extension, we have that $s' + v'$ is r_{n+1}, the T quadra-extension of r_n. So $v' \leq r_T$. By hypothesis, v' is discrete from both v and u', and so from $v + u' = u$. With Dichotomy (Axiom 2.7b), it is straightforward to check that v' is between u and t. But this contradicts the fact that u and t are quadra-extensions of each other. ■

Theorem 2.3. *Let r be any rectangle. Let R be the minimal closure of $\{r\}$ under quadra-extension, and let r^* be the fusion of R. Then r^* is G^2, the entire space.*

Proof Sketch: Suppose not. Then, since obviously $r^* \leq G^2$, $G^2 - r^*$ exists. So let m be any region that is discrete from r^*.

By Axiom 2.9, Unity of Space, there is a gq s such that (i) either $s = r$, or else both s is congruent with r and there is a direction T_1 of s such that r is T_1 of s and Aligned with s; and (ii) there is a direction T_2 of s, such that m overlaps the fusion of s and all gq's that are T_2 of s.

If $s = r$ then m overlaps the fusion of all gq's that are T_2 of r. Let r_1 be the T_2 quadra-extension of r. We have that m overlaps the fusion of r and all gq's that are T_2 of r. By Theorem 2.2, m overlaps the fusion of R and the fusion of the minimal closure of $\{r_1\}$ under T_2 quadra-extension. Either way, m overlaps r^*, which is a contradiction.

If, instead, $s \neq r$, then, by the definition of Alignment, there is a direction T of r such that s is T from r. As in Theorem 2.2, let r_1 be the T quadra-extension of r; and let R_T be the minimal closure of $\{r_1\}$ under the operation of taking the T quadra-extension. By a construction similar to that used in the proof of Theorem 2.2, there is a member r_n of R_T such that $s \leq r_n + r_{n+1}$, where r_{n+1} is the T quadra-extension of r_n. We have that there is a direction T_2 of s such that m overlaps the fusion of all gq's that are T_2 of s. It is straightforward to check that there is a direction T' of $r_n + r_{n+1}$ such that m overlaps the fusion of all gq's that are T' of $r_n + r_{n+1}$. Applying Theorem 2.2 again, we have that m overlaps the fusion of the minimal closure of $\{r_n + r_{n+1}\}$ under T' quadra-section. It is straightforward to verify that m thus overlaps r^*, a contradiction. ■

5.3 Generalization of the Archimedean Property and Recovery of the Parallels Postulate

So far, we have derived the Archimedean property as it pertains to any of the four directions relative to a given gq. There is, however, nothing so far that guarantees that gq's exist in arbitrary orientations, as usually understood in terms of lines or classes of parallel lines. In fact, the following describes a model $\mathfrak{M}$ of the axioms thus far. The "regions" of $\mathfrak{M}$ are regular, open sets of pairs of real numbers, with the usual metric and topology.[9] So the entire space G^2 of $\mathfrak{M}$ is $\mathbb{R}^2$. Mereological "parts" are non-empty subsets, and the fusion of a set S of regions is the interior of the closure of the union of the members of S. Let i be any open parallelogram, say one whose interior angles are $\frac{\pi}{4}$ and $\frac{3\pi}{4}$ (radians). Say that a regular, open set m is a gq of $\mathfrak{M}$ just in case m is a parallelogram whose sides are parallel to those of i. So, in $\mathfrak{M}$ all gq's are oriented the same way (i.e. they are all oriented the same as i). Let s and r be two regions of $\mathfrak{M}$. Say that $Cong(s, r)$ holds in $\mathfrak{M}$ just in case either s and r are gq's and s is congruent (in $\mathbb{R}^2$) to r, or else neither is a gq and $s = r$ (see note 4). It is straightforward to verify that $\mathfrak{M}$ satisfies all of the axioms of the previous section. In $\mathfrak{M}$ all gq's are "rectangles" (even though they are not rectangles in $\mathbb{R}^2$).

In effect, the Unity of Space axiom 2.9 and the principle 1.2 of unrestricted fusions guarantees that, in any model, there will be *regions* with all sorts of sizes, shapes, and orientations. For example, there will be regions that are *intuitively* "congruent" to a

[9] So $\mathfrak{M}$ is an analogue of one of the topological models of our one-dimensional theory (see Chapter 2).

given gq, but oriented differently, say at a $\frac{\pi}{6}$ angle from it. But, as indicated by the model $\mathfrak{M}$, we do not yet have the means to show that these regions *are* gq's and that they stand in the *Cong* relation to any gq's (despite axiom 2.2).

In this section, we add some further primitives on regions and axioms governing them that enable us to rule out models like $\mathfrak{M}$, and, indeed, to establish the full Archimedean property for arbitrary directions. These primitives, it turns out, also suffice for the full, regions-based recovery of Euclidean geometry, to be shown explicitly below in the case of the parallels postulate ("EPP").

The bulk of our task is definitional, introducing enough conceptual machinery to state axioms adequate for generalizing the Archimedean property and recovering Euclidean geometry. Our method employs a notion of "*angle*" as a region, written $Ang(\theta)$. In ordinary terms, we may think of an angle as a sector or a circle, or better, as an equivalence class of such sharing a common vertex, with the angles respectively of concentric circles such that they all share a common vertex and each is an "initial part describing the same angle", i.e. all are angles as sectors of larger concentric circles, all of the same angle norm (to be defined, below). But this is just heuristics. We will often drop the primitive '*Ang*', reserving θ, φ, ψ, etc. as angle variables (typically for generic representatives of equivalence classes). Of course, we don't yet have vertices as points, so we will need some new primitives on such regions and axioms governing them. It helps that we already have gq's, intuitively containing four angles (as regions) at "corners". Further, we have the resources to define explicitly the "corners" of a given gq. For example, using our relative direction terms, the "top-left corner" of given gq, r, can be identified with the nested sequence $\langle r_i \rangle$ of gq's r_i under iterated quadrasection starting with $r_0 = r$, setting $r_{i+1} =$ the top-left quadrasection of r_i (and similarly for the other three corners).[10] Thus we can simply identify the "vertex" of an angle ($< \pi$ radians) with the equivalence class of co-convergent n- and q-sequences defining the relevant "corner" of a gq from which it derives. Further, since angles in our sense are regions, we already have applicable our primitive congruence relation, which, in the case of angles, will be written $Cong(\theta, \varphi)$, where, as usual, this means, in ordinary terms, "same size and same shape". Here are the axioms we will use:

Axiom 3.1a: Let r be a gq; then r has exactly four angles, $Ang(\theta_i), i = 1, \ldots, 4$, with each $\theta_i < r$, such that the vertex of $\theta_1 =$ the *UL* corner as defined, and likewise proceeding clockwise. Further, diagonally opposite angles are congruent, $Cong(\theta_1, \theta_3)$ and $Cong(\theta_2, \theta_4)$.

[10] The definition of '*point*' in our framework is given below via properly nested sequences of gq's under iterated nona-sectioning, and co-convergence of these with properly nested sequences under quadra-sectioning is also defined.

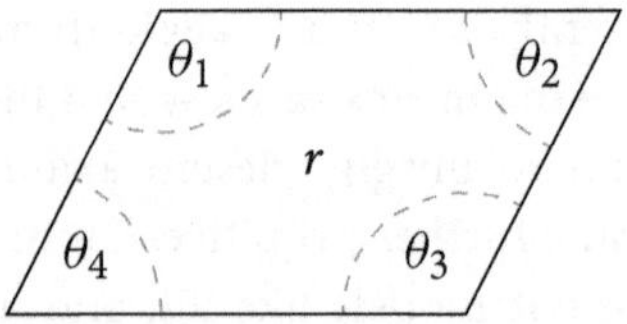

We also want to stipulate that an angle, as part of a gq "fills up" a corner of the gq in the sense of having "sides" that coincide with segments of two adjacent sides of the gq. This can be expressed using the relative direction terms, "to the *D* of", as in the previous section:

Axiom 3.1b: Let r be a gq and let θ be an angle at the FF′ corner of r (where 'FF′' takes the values 'UL', 'UR', 'DL', or 'DR'); then no part of θ lies F of r or F′ of r, and no part of r lies F of θ or F′ of θ.

Based on this, we can redefine *rectangle(r)* as: r is a gq all four of whose angles are congruent.[11] Then each such angle, θ, is called "*right*", and its norm, $|\theta| = \frac{\pi}{2}$. By our axiom (2.13) of the previous section, rectangles exist (in abundance).

Given just this much, we already can *define* a suitable relation of *adjacency* of two angles to one another:

$$Adj(\varphi, \psi) \text{ iff } \varphi|\psi \;\&\; Ang(\varphi + \psi).$$

It is convenient also to introduce two subrelations of angle adjacency, $+Adj(\varphi, \psi)$ and $-Adj(\varphi, \psi)$, the former defined as $Adj(\varphi, \psi)$ & φ is clockwise of ψ, the latter as the same except substituting 'counterclockwise' for 'clockwise'.

Invoking quadraextensions, we have that each right angle, say, θ_1, of a rectangle is adjacent to two others (one, θ_2, $+Adj$ to θ_1, the other, θ_3, $-Adj$), and those two are adjacent to a fourth, θ_4, completing a *circle* (as a region) $= \Sigma_{i=1}^{4}\theta_i$. All circles may be identified in this way, as an angle.

Our next axiom is one of Euclid's:

Axiom 3.1c: Let θ and ψ be any two right angles; then $Cong(\theta, \psi)$.

If C is a circle, its norm, $|C| = 2\pi$.

In general, the (angle) norm function is stipulated to be countably additive with respect to sequentially adjacent angles: $|\Sigma_j\theta_j| = \Sigma_j\,|\theta_j,|\;\bmod 2\pi$, whenever $Adj(\theta_j, \theta_{j+1})$, for each j, with $|\theta_j| \leq 2\pi$ (where we have abused notation, using the Σ inside the brackets to denote mereological fusion and the Σ outside to be ordinary addition of real numbers). This is taken to apply to finite sums as well as countably infinite ones.

[11] As indicated by the above model $\mathfrak{M}$, this definition of "rectangle" is not equivalent to that of the previous section. It is straightforward that rectangles, in the present sense, meet the defintion of "rectangle" in the previous section.

Axiom 3.2 Every angle has a (polar) bisection: $\forall\theta\exists\varphi, \psi[Cong(\varphi, \psi)$ & $Adj(\varphi, \psi)$ & $\theta = \varphi + \psi]$ ('+' in the sense of mereology). Except in the case of θ = a circle, the bisection is unique.

This already implies that angles are "bilaterally symmetric", and hence that, as it is usually expressed in terms of points and lines, the two "sides" of an angle (as a circle sector) are "of equal length". And since Axiom 3.2 can be iterated any finite number of times, this guarantees that all of the angles obtained under iterated polar bisection have sides of the same length as one another. Our next axiom enables us to extend this to subangles of arbitrary real norm (mod 2π) relative to that of any given angle.

Axiom 3.3 Let $\langle\theta_j\rangle$ be a sequence of angles, each $\theta_j < \theta$, such that $Adj(\theta_j, \theta_{j+1})$, for each j; then the fusion, $\Sigma_j\theta_j$, of the θ_j is an angle, θ^j, with $\theta^j \leq \theta$.

Angles of a sequence $\langle\theta_j\rangle$ satisfying the hypothesis of this last axiom will (collectively) be called "sequentially adjacent". If all the *Adj* instances are in the same sense, + or −, we call the angles "sequentially +adjacent" and "sequentially −adjacent", respectively. Since angles of irrational norm (mod 2π) can be obtained exactly by fusion of convergent sequentially adjacent angles of binary rational norm, the effect of these last two axioms is to furnish us with a continuum of angles of any given circle, in one-one correspondence with those as usually defined via intersecting half-lines.

Our next axiom ensures that fusions of sequentially adjacent angles preserve congruence:

Axiom 3.4 If the angles of $\langle\theta_j\rangle$ are either sequentially +adjacent or −adjacent and those of $\langle\varphi_j\rangle$ are either sequentially +adjacent or −adjacent, with all the $\theta_j < \theta$ and all the $\varphi_j < \varphi$, and $Cong(\theta_j, \varphi_j)$, for all j, then $Cong(\Sigma_j\theta_j, \Sigma_j\varphi_j)$.

Similarly we adopt a kind of "converse" to this:

Axiom 3.5 Under the same hypotheses as the previous axiom but with $Cong(\theta_j, \varphi_j)$ holding for all j except for one, $j = k$, then if also $Cong(\Sigma_j\theta_j, \Sigma_j\varphi_j)$, then $Cong(\theta_k, \varphi_k)$.

Our next step is to obtain the effect of rotation operators, here understood as rotating a given angle through another given angle. Once that is accomplished, we will be able to state an axiom guaranteeing the existence of gq's (or rectangles) oriented in arbitrary directions in space. Then the derivation of the Archimedean property of the previous section will straightforwardly generalize as desired.

In the usual terms of points and lines, any angle (in our sense of region, as well as on the standard conception) uniquely determines a circle, with the vertex as center and either side of the angle as radius. How can we express this function from angles to circles just in terms of regions? One requirement for a given angle θ to be *of* a circle C is that the fusion of θ together with sufficiently (finitely) many copies of θ, sequentially adjacent in the same sense (+ or −), is identical with C.[12] Thus we postulate

[12] It makes no difference if the last copy of θ returns to θ with an overlapping excess: the fusion of the sequence is still just the circle determined by θ.

Axiom 3.6 *(Angle biextensions)* For any given angle θ ($< 2\pi$), there are exactly two unique angles, θ', θ'', such that $Cong(\theta, \theta')$, $Cong(\theta, \theta'')$, and $+Adj(\theta', \theta)$ and $-Adj(\theta'', \theta)$.

Then we define θ *is of* C to mean that, for some n, there are n angles, θ_j, each congruent to θ, sequentially +adjacent (or −adjacent), with $\theta_1 = \theta$, such that the fusion, $\Sigma_j \theta_j = C$. (Of course, then $\theta \leq C$.)

Now let C and C' be given congruent circles, and let θ be an angle of C, and let φ be an angle of C'. Our next axiom generalizes the last, guaranteeing that copies of either angle exist adjacent to either side of the other.

Axiom 3.7 *(Angle translation)* Given circles C, C' with $Cong(C, C')$ and angles θ of C and φ of C', each less than 2π, there are unique φ', φ'' with $Cong(\varphi', \varphi)$ and $Cong(\varphi'', \varphi)$ and $+Adj(\varphi', \theta)$ and $-Adj(\varphi'', \theta)$.

(Since the circle and angle variables in this are universally quantified, it already implies the statement with the roles of θ, φ, and C, C' reversed.)[13]

Angle translation and the intended interpretation of *Adj* also provide a route to recovering the usual definition of 'circle' via equal-length radii from a fixed point (center, vertex of angles in our sense).

The pieces are now in place for us to define general rotation operators on angles, and hence on gq's. Specifically, for example, we would like to complete a formula that defines

$+R(\theta, \theta', \varphi)$ iff θ', with $Cong(\theta, \theta')$, represents "the result of rotating θ clockwise through angle φ", where $+Adj(\varphi, \theta)$.

This can be done without defining any new relations as follows: first, apply biextension to φ, obtaining φ' with $Cong(\varphi', \varphi)$ and $+Adj(\varphi', \varphi)$; then require that $-Adj(\theta', \varphi')$. Now, in terms of line segments, the same sides of θ and θ' form angles congruent to φ. Formally, we first define the four-place relation

$+R(\theta, \theta', \varphi, \varphi')$ iff $Cong(\theta, \theta')$ & $+Adj(\varphi, \theta)$ & $Cong(\varphi, \varphi')$ & $+Adj(\varphi', \varphi)$ & $-Adj(\theta', \varphi')$.

Then we define

$+R(\theta, \theta', \varphi)$ iff $\exists \varphi'[+R(\theta, \theta', \varphi, \varphi')]$.

'$-R(\theta, \theta', \varphi)$', "$\theta'$ results from counterclockwise rotation of θ through φ", is defined analogously.

The last axiom of this group is key to proving our main results of this and the next section, the generalization of the Archimedean property to arbitrary directions and recovering the EPP:

[13] This axiom also implies the previous one on angle biextensions as a special case. The latter, however, is needed in order to infer that the definition of "θ *is of* C" is instantiated, as in the axiom of angle translation.

Axiom 3.8 *(Angles to gq's)* Let θ be any angle $< \pi$; then there is a gq, r, with θ as one of its angles.[14]

As a special case, let r be a given rectangle with ψ as one of its right angles, and let φ be an arbitrary angle $< \pi$; then, by Angle translation, there exists ψ' with $Cong(\psi, \psi')$ such that, for some φ' satisfying $Cong(\varphi, \varphi')$, we have that $+R(\psi, \psi', \varphi, \varphi')$ (and similarly for $-R$), i.e. ψ' results from positive (negative) rotation of ψ through angle φ. Then by the "Angles to gq's" axiom, there is a rectangle r' with ψ' as one of its angles. Since φ is arbitrary, this furnishes us with rectangles with arbitrary orientations, as desired. Thus, we have:

Theorem 3.1. Same statement as Theorem 2.2: *Let r be any rectangle and let T be any direction from r. Let r^T be the fusion of all regions that are T of r. Let r_1 be the T quadra-extension of r, and let r_T be the minimal closure of $\{r_1\}$ under the operation of taking the T quadra-extension. Let r_T be the fusion of r_T. Then $r_T = r^T$.*

Here, however, the variable 'r' over rectangles ranges over those of arbitrary orientations. The proof of Theorem 2.2 carries over intact. Similarly with regard to the next theorem and its proof:

Theorem 3.2. Same statement as Theorem 2.3: *Let r be any rectangle. Let R be the minimal closure of $\{r\}$ under quadra-extension, and let r^* be the fusion of R. Then r^* is G^2, the entire space.*

Now we turn to the recovery of the EPP. This breaks down naturally into two tasks: first, we need to introduce '*point*' and '*line*' and enough relations among these to be able to *express* the EPP in our framework. Second, we then will introduce '*triangle*' and the relation of '*similarity*' between triangles and some axioms governing these predicates as a means of recovering the Euclidean proof of *uniqueness* of parallels (through a given point, in relation to a given line). The *existence* of parallels already will follow from our arbitrarily oriented gq's.

Turning to the first task: We will want to introduce "lines" as certain sets or pluralities of points, where points are taken to be sequences of properly nested gq's. For this, we must be assured of distinct points, which translates as "non-co-convergent" sequences. Thus, we need a criterion of co-convergence (intuitively meaning, "to the same point").

Note that two co-convergent sequences (speaking informally here), if obtained by repeated *quadra*-secting, may be entirely discrete from one another, i.e. no region of one sequence overlaps any of the other. It can even be that no region of one is adjacent to any of the other. (For example, take one sequence to be that defined in the proof of

[14] Although angles were first introduced above as deriving from gq's, occupying their "corners", this axiom is not redundant as it applies to arbitrary angles, including all those obtained from given ones via the operations of bisection, biextension, translation, and fusion of sequentially adjacent angles.

Theorem 2.1, above, and take a second, also converging to the "center" of r, to be that which results from switching the roles of 'NW' and 'SE' in the definition of the first sequence.) However, if we restrict our definition to sequences of gq-regions obtained always by *nona-secting* the previous region, then the criterion for co-convergence is especially simple, and, moreover, the proof that non-co-convergent sequences exist is made easy. So we define:

A properly nested sequence $\sigma = \langle \sigma_i \rangle$ of gq-regions is an *n-sequence* iff each σ_{i+1} is the central region of the nona-section of σ_i.

Properly nested n-sequences $\rho = \langle \rho_i \rangle$, $\sigma = \langle \sigma_j \rangle$ are *co-convergent—CoConv* (ρ, σ)—iff

$$\forall i \exists j[\sigma_j \leq \rho_i] \ \& \ \forall j \exists i(\rho_i \leq \sigma_j).$$

It is easily proved that *CoConv* is an equivalence relation. (This is left to the reader.)

Now we prove that there are many non-co-convergent n-sequences.

Lemma 3.3. *Given any* n-*sequence* σ, *there exists an n-sequence* τ *such that* $\neg CoConv(\sigma, \tau)$.

Proof. If the first term σ_1 of σ is part of gq-region r, let τ_1 be any gq-region such that $\tau_1 < r'$, where r' can be any region resulting from applying quadra-extension to r. Then, since τ_1 is discrete from σ_1, in fact all the τ_i are discrete from all the σ_j by the choice of central regions of nona-sections at each stage. ■

In light of our previous axioms, this provides many non-co-convergent pairs of convergent sequences, hence many mutually distinct "points" and, as will emerge, distinct "lines" as well.

It will also be useful to invoke properly nested gq-sequences obtained by *quadra-section* at each stage, especially when speaking of "corners" and "border points" of gq-regions (cf. Axiom 3.1a, above). Such points must be defined by n-sequences, but we may say that properly nested sequences via quadra-section—call them "*q-sequences*"—are also co-convergent with a fixed n-sequence. For this, we may define

Q-sequences ρ, σ *are co-convergent with n-sequence* τ*—CoConv*(ρ, σ, τ)—iff $\forall\ i, j, k$ $[\ \tau_i \circ \rho_j \ \ \& \ \tau_i \circ \sigma_k]$. Holding n-sequence τ fixed, this is also an equivalence relation between the first two terms, provided the relation holds (i.e. reflexivity, *CoConv*(σ, σ, τ), obtains whenever the condition $\tau_i \circ \sigma_j$ for all i, j, is met).[15]

We define "*p is a point*" to mean p is an equivalence class of properly nested co-convergent n-sequences. Occasionally, when convenient, we will refer to a generic member of such a class rather than to the class itself.

[15] What we call n-sequences are closely related to a special case of "representatives of points", as introduced in [51]. See also [18]. The relation of co-convergence is analogous to the relation of mutual covering of sets of regions. Much of this work is inspired by [128].

Our definition of line, one of several equivalents chosen for convenience, follows closely that of classical analysis applied to geometry. Stated informally, "*l is a line*" iff l is a maximal class of "*collinear*" points, where three distinct points—i.e. pairwise non-co-convergent—n, p, q, are defined to be *collinear*—$Coll(n, p, q)$—just in case they are all points of a common "border" of a gq, r. This in turn is spelled out via q-sequences co-convergent with the n-sequences defining the respective points. For example, to be a point of, say, the U ("up") border of gq r is to be specified by a q-sequence $\langle \sigma_j \rangle$, where $\sigma_1 = r$ and $\sigma_{j+1} =$ the "UX-quadrasection of σ_j", where 'X' can vary over relative directions R and L. Similar patterns specify the other "border points" of r, *mutatis mutandis.*

Our next two lemmas establish the equivalence of our definition of 'line' with the more common one as a pair of distinct points.

Lemma 3.4. *Let p and q be distinct points; then they both lie on exactly one line.*

Proof. Invoking rectangles of an n-sequence for p, say, and rotation of one of these if needed, one obtains a rectangle r with p and q as adjacent corners. One of the borders, call it B, of r connects p and q.

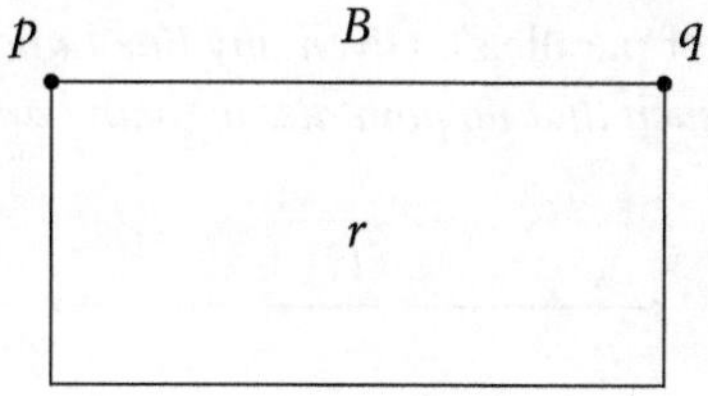

This determines a line l thus: take the minimal closure of r under biextension in the two directions orthogonal to B; l then consists of all the points of the B-borders of these rectangles. Clearly it is a collinearity class, and the Archimedean property ensures that it is maximal. To prove the uniqueness of l, suppose both p and q also lie on a line $k \neq l$. Since k and l are not parallel, they meet at a point, m, which may be supposed to be p or q or neither. Now it is a property of gq borders that any two points of a border are at a straight angle (of norm π) to each other (the vertex of which is the mid-point between them). (Cf. Lemma 3.7, below.) Suppose m is p; then there is a point n of k such that p is equidistant from q and n (via a gq with segment $\overline{pn}$ of k as a border congruent to a gq with a segment $\overline{pq}$ of l as a border). But the angle qpn formed at $p \neq \pi$ unless the segment $\overline{pn}$ is of line l as well as k. In that case, by an induction, iterating this construction along k indefinitely in both directions would show that $k = l$, contradicting the hypothesis. Thus, we can suppose that $\overline{pn}$ is of k but not of l, and q, p, n are not collinear with n, therefore p and q cannot both lie on k, contradicting the hypothesis. The other two cases, assuming $m = q$ or m distinct from p and q, are argued analogously. This proves the uniqueness of l as the maximal collinearity class determined by the pair, p, q. ■

Corollary. *Two distinct non-parallel lines share exactly one point in common.*

Proof. Immediate from Lemma 3.4. ■

We now have a kind of "transitivity" of the collinearity relation:

Lemma 3.5. *If* $Coll(n, p, q)$ *and* $Coll(p, q, u)$, *then both* $Coll(n, p, u)$ *and* $Coll(n, q, u)$.[16]

Proof. By definition, the hypothesis together with the previous lemma implies that n, p, q, and u all lie on the same line, l, viz. that uniquely determined by the pair p, q, whence the conclusion follows by definition of the *Coll* relation. ■

For any gq, r, we have the minimal closure, $Cl^{X,Y}(r)$, of $\{r\}$ under biextension in two given opposite directions (i.e. $X = \text{U}$, $Y = \text{D}$ or $X = \text{L}$, $Y = \text{R}$). Lines in our sense can be thought of as the common bi-infinite borders of the fusion of all the gq's in any one of these minimal closures. It should be clear that these constructions along with Axiom 3.8 on the "isotropy" of angles and their parent gq's, guarantee that our space has a plenum of lines in 1-1 correspondence with those of ordinary point-based Euclidean geometry, or classical (Cartesian) analytic geometry.[17]

This much already suffices to recover half of the EPP. Here is an especially easy proof.

Theorem 3.6. (Existence of parallels): *Given any line l and a point p not on l, there exists a line l' with p on l' such that no point lies on both l and l'.*

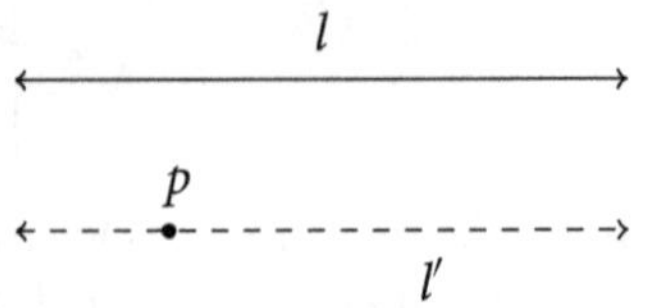

Proof (sketch): The idea is to start with an arbitrary gq r with a segment b of l as one of the borders of r. (By the definition of '*line*', there is, for any point q on l, such an r with q on a border of r.) Without loss, suppose that p is D and D′ of r, where D and D′ are two adjacent directions relative to r, e.g. U and R, and where one of these, say it's U, is opposite to the border b. (In favorable cases, one of the directions would suffice, or p could already be "in r", where that means that, for any n-sequence $\sigma_p = \langle\sigma_i\rangle_p$ defining p, for some j and every $m \geq j$, $\sigma_m \leq r$.) Next extend r in these directions a finite number of times in each which are sufficient to result via fusion in a gq $r' \geq r$ so that p is in r'. (Here the Archimedean property of our space is used.) Call that fusion

[16] In this and several further lemmas and theorems, we are using vocabulary of point-based geometry. But also such terms are used as defined in our framework of regions, and, furthermore, they do not occur, even as defined, in our axioms (contrast this with [119]). Thus, our procedures are entirely in line with those of a full-fledged reduction of one theory to another.

[17] In effect, we have just sketched a key part of the recovery of point-based geometry as superstructure over our regions-based space.

F. If p already is a U-border point of r', we're done, as the bi-infinite L–R extension of F then has as its U border a parallel, l', to l. In the general case, it is necessary to subtract a gq from an end of F (from the "top" of F in the case where b is the D (down) border of F). (This is always possible by sufficiently finely subdividing F, via quadrasection along the axis determined by U, and taking fusions of Cauchy sequences of gq's (in this instance, strips extending from L to R) as needed.) The result of such a subtraction is then a gq, F', with p lying on its U border. The bi-infinite extension of this then provides the desired parallel l'. The proof that l' so constructed never meets l invokes the fact that any n-sequence defining a point of l' is eventually discrete from any such n-sequence defining a point of l (in accordance with visualization). ■

To recover the uniqueness of parallels, a bit more apparatus is useful, viz. the introduction of *triangles* and the essentially Euclidean phenomenon of ever larger *similar* triangles to a given one. (Indeed, the existence of similars that are non-congruent is just a version of the EPP, inter-derivable with the Playfair version we're recovering. The absence of non-congruent similars is, of course, one of the key elementary features of spaces of non-zero curvature.)

We define "*triangle*" as a diagonal half of a gq: $Tr(t)$ iff $\exists r, t'[r$ a gq & $r = t + t'$ & "t, t' result from diagonal bisection of r"], where the clause in quotes is defined thus: t shares two adjacent borders with r and t' shares the other two adjacent borders with r & $Cong(t, t')$ & $Adj(t, t')$, where a border of a gq is defined via q-sequences as above, and adjacency of such borders is defined via the disjunction of relevant pairs of border-labels, proceeding clockwise, i.e. (U,R) $\vee$ (R,D) $\vee$ (D,L) $\vee$ (L,U).

To ensure existence of triangles, we add to the group of axioms labeled 3.1 the following:

Axiom 3.1d Let r be a gq region; then r has two diagonal bisections, each into two congruent, non-overlapping triangles, t, t' (with $r = t + t'$), each having exactly three angles, θ, φ, ψ of t, θ', φ', ψ' of t', with $Cong(\theta, \theta')$, $Cong(\varphi, \varphi')$, and $Cong(\psi, \psi')$, and satisfying $Adj(\varphi, \psi')$ and $Adj(\psi, \varphi')$ with $\varphi + \psi'$ an angle of r congruent to $\psi + \varphi'$, the diagonally opposite angle of r, and with θ and θ' the remaining two angles of r, respectively.

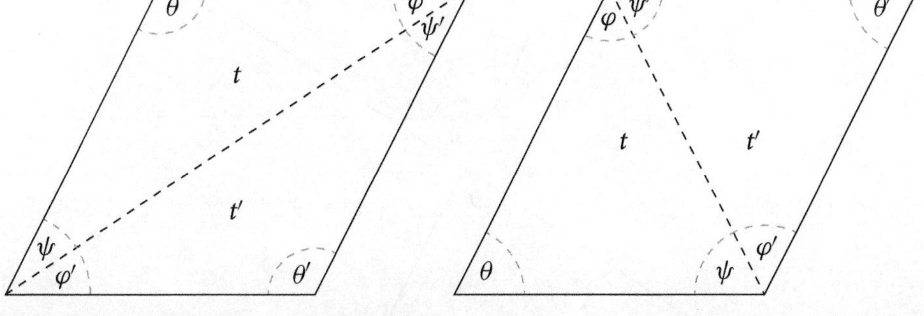

In view of our axioms on angles and gq's, this guarantees arbitrary triangles existing everywhere and in all orientations ("ubiquity" and "isotropy" of triangles, inherited from those properties of gq's).

Now suppose, as in the hypothesis of the statement of the uniqueness of parallels, we are given a line, l, and a point, p, not on l. Suppose, further that we have available also a line, l', through p and parallel to l, as already established. The proof should then proceed to demonstrate that any other line, k, through p eventually meets the given l. That will follow by constructing ever larger similar triangles (say, by doubling segments) formed by segments of l' and segments of k, both proceeding from p (as a vertex common to all the triangles involved) in the direction D or D− such that that half of k "lies on the l-side of l'". Of course, we need to be able to render this in our regions-only language. And we need to be sure that our "lines" as defined behave in accordance with the presupposition, that lines in the roles of l' and k indeed "cross one another" at p so that there are indeed a half-line of k and a direction D such that, to the D (or D−) of p, k lies on the l-side of l'.

A couple of definitions:

Given a line l and points p and q of l, point m is *between* p and q, *Betw*(p, m, q), just in case p and q are corners of a border, b, of a given gq, r, such that $m \neq p$, $m \neq q$, and m is also a point of b.

Let l be a line; then a (finite) *segment s of l* is the set or plurality of all points between any two given points, p and q, of l.

The next lemma states a fact cited above in the proof of Lemma 3.4:

Lemma 3.7. *Let l be a given line with p an arbitrary point on l. Then p is the vertex of a straight angle (of norm π) formed by a segment of l about p.*

Proof. Since p is defined by an n-sequence, indeed of rectangles, it is also a "corner" of four rectangles, meeting at p as the center of a larger rectangle. By a simultaneous rotation of these rectangles, the same border of two of them adjacent to each other, we obtain a segment of l about p as desired. ■

Lemma 3.8. *Let lines l and l′ meet at p; then segments of them form four angles at p such that the two opposite angles of either of the two pairs of opposite angles are congruent to one another.*

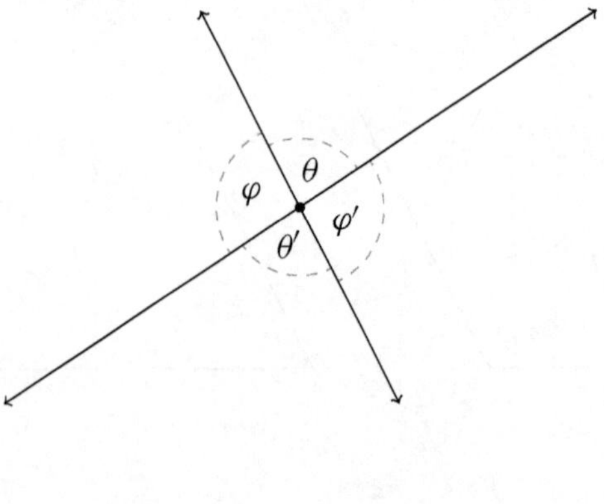

The proof employs Lemma 3.7 repeatedly and is just the familiar one of standard Euclidean geometry, and it is left to the reader.

We will also need an axiom to ensure the obvious requirements that congruence of two gq's suffices for congruence of the triangles of one with those of the other, and that congruence of two gq's or of two triangles implies congruence of corresponding angles of those respective figures.

Axiom 3.9a Let $Cong(r, r')$ for gq's r and r', and let t and t' be triangles of r and r', respectively; then $Cong(t, t')$.

Axiom 3.9b Let $Cong(r, r')$ for gq's r and r' and let the four angles θ_i of r correspond to the four angles θ_i' of r', as indicated by their indices; then $Cong(\theta_i, \theta_i')$, for each i.

Axiom 3.9c Let $Cong(t, t')$ for triangles t and t' and let the three angles θ_i of t correspond to the three angles θ_i' of t', as indicated by their indices; then $Cong(\theta_i, \theta_i')$, for each i.

The pieces are now in place to prove the uniqueness part of the EPP:

Theorem 3.10. (Uniqueness of parallels): *Given a line l and a point p not on l, there is a unique line l′ through p that is parallel to l.*

Proof. Assume, for a contradiction, that, in addition to a parallel, l', to l through p (constructed as in the proof of Theorem 3.5), there is a second parallel, k, to l through p. By Lemmas 3.6, 3.7, and 3.8, k and l' form an angle $\theta < \pi$ on the l side of l'. For ease of exposition, suppose that l is "down" relative to the gq's defining l', and that k proceeds "down" and "left" relative to those gq's. Further, we may assume, without loss, that those gq's are rectangles $\langle r_j \rangle, j = \ldots -2, -1, 0, 1, 2, \ldots$, and that it is their "up" borders that define l' (so that they extend toward l). Either one of those rectangles (say r_0) or a rectangle that is a proper part of one of them has point p as its UR corner. Let q_0 be any other point of k down and left of p, chosen so that q_0 lies "up" from given line l.

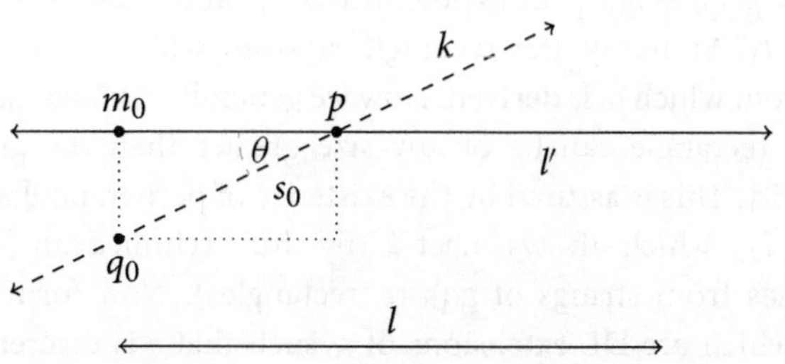

There is a rectangle, s_0, with p as its UR corner, q_0 as its DL corner, with its UL corner a point m_0 on l', and with its DR corner a point p_0 on a line $l_\perp$ through p perpendicular to l' and l. (By bisecting the straight angle centered on p, licensed by Axiom 3.2, a perpendicular to a given line at one of its points always exists and is unique. That a line perpendicular to both l' and l exists is guaranteed by the construction of parallels, as in the proof of Theorem 3.5, where the gq's involved are taken to be rectangles.) By Axiom 3.1d, there is a triangle, t_0, derived from s_0 by diagonal bisection, with corners (vertices) p, q_0, m_0. Now, quadra-extending s_0 down and left and quadra-extending the left one down, construct a larger rectangle, s_1, with s_0 forming its UR quadrant. In addition to p as UR corner of s_1, designate the other corners p_1 (DR), q_1 (DL) and m_1 (UL). Then the diagonal bisection of s_1 defined by a segment of k (from p to q_1), yields as upper triangle t_1 containing t_0 as a proper "initial" part (proceeding down and left from p).

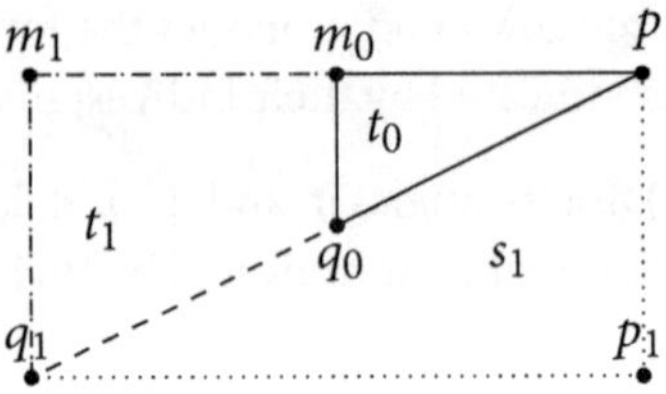

Claim: Triangles t_0 and t_1 are similar. Proof of Claim: By Axiom 2.1, the four rectangles making up s_1 are congruent; so, in particular, the two pairs of triangles formed by k as constituting the UR quadrant and the DL quadrant of s_1 are all congruent, whence—by Axiom 3.9c—so are their corresponding angles, in particular the two formed by k and the "verticals" perpendicular to l', proving the claim. It follows that the respective borders of t_1, in particular the verticals just mentioned, are twice the length of those of t_0. (This can be taken as a definition of relative lengths of line segments.) Since, by the Archimedean property, the vertical distance between l' and l is finite, a finite number of doublings of rectangles, of sequence $\langle s_i \rangle$, and their (upper) triangles, $\langle t_i \rangle$, suffices to ensure that some t_i overlaps one of the rectangles whose U borders serve to define the given line l.[18] The existence of an intersection point of k and l can now be established as follows. Call the triangle t_1 defined above a "DL-*extension*" of t_0, i.e. with t_0 similar to t_1 and with segment $\overline{p, q_1} = \overline{p, q_0} + \overline{q_0, q_1}$ as hypotenuse of t_1 (and likewise, *mutatis mutandis*, for the arms of t_1. Similarly, the rectangle s_1 from which t_1 is derived is a DL-extension of s_0 from which t_0 is derived. Now we generalize this so that DL-extensions of a triangle or rectangle can be of any size greater than the given figures, but retaining similarity. This is assured by the existence of perpendiculars from arbitrary points of l' and $l_\perp$, which always meet k (by the Archimedean property and the derivation of lines from strings of gq's or rectangles). Now form the fusion F of all rectangles s which are DL-extensions of s_0 such that s is discrete from the half-space—call it H_D—down from given line l. We claim that F is a rectangle whose D border is a segment n of l, and hence a point, call it q^*, of which is also a

[18] In a point-based analytic treatment, an analogue of the intermediate value theorem would establish that line k meets line l. That argument, of course, is not yet available in our framework, so we still must establish an intersection point of k and l.

point of k—in fact the point being the DL corner of F. Suppose not: there are two cases: Case (1) F overlaps H_D. This is impossible, since by mereology, if a fusion of objects meeting a condition C overlaps an object x, then at least one of the C objects overlaps x, and, in this case, C requires discreteness from H_D. Case (2) Some region r is down from F but up from H_D (in this sense, "lying between" F and H_D. But this too is impossible, since then there would be a DL-extension of s_0 beyond F and overlapping r yet still discrete from H_D, contradicting the hypothesis that F is the fusion of all DL-extensions of s_0 discrete from H_D. That such a DL-extension of F exists can be seen by taking a rectangular part of r, guaranteed by Axiom 10 of the previous section, reorienting it (via a rotation) if necessary so that its Up and Down borders are parallel with l, and extending it Left so the DL corner of the extended rectangle meets k, and extending it Right to meet $l_\perp$ as needed to construct a DL-extension of F, contradicting the definition of F. Thus, F is indeed adjacent to a rectangular strip of H_D bounded above by the D-border of F, i.e. that border of F must indeed coincide with a segment of l, which we labeled n. Furthermore, the L-border of F does not overlap the half-space H_L defined as follows: let q be any point of k lying within the half-space H_D (i.e. such that any n-sequence for q eventually is part of H_D), and let v_q be the vertical line through q perpendicular to l'. Let H_{v_q} be the half-space left of v_q, and let H_L be the fusion of all H_{v_q} as q varies over the points of k lying in H_D. By the construction of F, it must be discrete from H_L. Furthermore, by an argument exactly analogous to that just given relating the D-border of F to H_D, it follows that the L-border of F is adjacent to a rectangular strip of H_L defined up and down by the parallels l and l', respectively. This enables a definition (via an n-sequence) of the DL corner of F, and this specifies a point q^* as the intersection point of l and k. ■

5.4 Closing Reflections

The above thus recapitulates the essentials of Euclidean geometry in a regions-based framework. In particular, the "points", defined above, are isomorphic to $\mathbb{R}^2$. As in the one dimensional case, it is also fairly straightforward to construct models of the regions-based theory (or theories) in the more usual punctiform $\mathbb{R}^2$. In one such model, the "regions" are regular, open sets of points, and the "gq's" are open parallelograms. In another model, the "regions" are regular closed sets of points, and the "gq's" are closed parallelograms. In both cases, a "region" m is part of a "region" m' if $m \subseteq m'$. The other primitives are straightforward.

It is also fairly straightforward to extend the two-dimensional theory to three or more dimensions. We'd begin with the notion of a "generalized rectangular solid (grs)" and go from there. For example, in the case of three-dimensional Euclidean space, Axiom 2.1 would be revised to require the existence of grs's, each having decompositions into eight mutually congruent, non-overlapping but "adjacent" grs's and into twenty-seven mutually congruent grs's (in place of quadra-sections and nona-sections, respectively), and having also six extensions by congruents, each meeting the given grs at a face, etc. In general, for $n > 3$, the corresponding axiom would specify, for each n-dimensional "hyper-grs", decompositions into 2^n and 3^n sub-hyper-grs's, respectively, and $2n$ extensions, one for each of the $2n$ directions determined by the n linearly independent coordinate axes. Similar generalizations would be laid down for axioms on angles. It should also be possible to develop various regions-based non-Euclidean geometries, the most challenging (and significant for relativistic physics) being those of variable curvature; but that must be left for future work.[19]

Appendix: Simplified Proof of Theorem 2.2

First, add two further axioms to those of 2.12, the first a variant of Translation:

Axiom 2.12f: Given a rectangle r and a direction D defined by r, and a region m D of r, there is a rectangle r' D of r, and Congruent to r, and Aligned with r such that $r' \circ m$.

Also, we want to express that rectangles are "connected" regions, intuitively lacking any gaps. One necessary (though not sufficient) condition of this goes as follows:

Axiom 2.12g: Given a (finite or countably infinite) sequence of rectangles $\langle r_j \rangle$ such that each beyond the first is the D quadra-extension of the preceding; and given another rectangle s, Aligned with the r_j, suppose for a particular j that $s \circ r_j$; suppose further than $s \circ m$ for some region m that is D of the sequence $\langle r_j \rangle$ or its fusion; then $r_k < s$ for all $k > j$.

The pieces are now in place to prove the following important theorem:

Theorem 2.2: Let r be any rectangle and let T be any direction from r. Let r^T be the fusion of all regions that are T of r. Let r_1 be the T quadra-extension of r, and let R_T be the minimal closure of $\{r_1\}$ under the operation of taking the T quadra-extension. Let r_T be the fusion of R_T. Then $r_T = r^T$.

Proof: Suppose $r_T \neq r^T$; then $r^T - r_T$ exists. Then by Axiom 2.12f above, there exists r' with r' T of r, *Cong*(r', r) and Aligned(r', r) such that $r' \circ r^T - r_T$. In fact, we have as a

Lemma: $r' < r^T - r_T$.

Proof. Suppose not. Then, as r' is T of r and hence $< r^T$, we have $r' \circ r_T$, and so it must overlap some r_i as a copy of the original r proceeding in the T direction. Let r_j be the least copy of r that r' overlaps. But $r' \circ r^T - r_T$. But this is, by definition T of r_T. Then taking $r^T - r_T$ as m in Axiom 2.12g, it follows that $r_n < r'$ for all $n > j$. But then we have that r' is not Congruent to r, by Axiom 2.12b (proper parts are never Cong to the whole). QED for the Lemma.

19 But see the discussion of differential geometry in Chapter 7.

Now apply the Lemma and the Translation Axiom to infer existence of a rectangle s such that Cong(s, r) and Aligned(s, r) such that $s < r^T - r_T$ but also s is Contiguous with r_T. But the latter is impossible: for let s' be the R^- quadra-extension of s (i.e. in the direction proceeding backwards toward the original given r). Then $s' \circ r_T$, by the Contiguity of s and r_T. But then there is a least j such that $s' \circ r_j$. But then, since s is the T quadra-extension of s', we must have *either* that $s \circ r_{j+1}$, which contradicts that $s < r^T - r_T$; *or* we have that $r_{j+1} < s'$; but this latter contradicts Axiom 2.12b. ■

6

Non-Euclidean Extensions

The purpose of this chapter is to show how to modify the regions-based, two-dimensional Euclidean space, developed in the previous chapter, to produce regions-based non-Euclidean spaces. The first section produces a hyperbolic geometry, one with negative constant curvature -1, and the second a spherical geometry, with constant positive curvature $+1$.

6.1 Hyperbolic Space

6.1.1 Essentials

In the standard, point-based treatments, a two-dimensional, hyperbolic space arises by adopting all of the Euclidean axioms for the plane (E^2) except the parallels postulate, which is replaced by the following:

> **Multi-Parallels Postulate ("MPP"):** *Given any line l and a point p not on l, there exist more than one line (or infinitely many lines), l', through p which never meet l (i.e. such that $l' \| l$).*

In the model, H^2, of these axioms, of interest here, there are infinitely many parallels. Two of these, relative to a given line l and point p, are asymptotic to l, in opposite directions, and are called "limiting-parallels", between which lie all the other parallels. Furthermore, the angle sums of polygons are less than their Euclidean counterparts, with the difference ("defect") shrinking as the polygons considered are smaller and smaller, approaching, but never reaching, the Euclidean values. In particular, this implies that, in H^2, any polygon similar to a given one (i.e. corresponding angles are equal) is also congruent to it. Thus, we see that our Axiom 2.1 on quadra-sections, from the previous chapter, already builds in the EPP, and must be revised in a regions-based construction of H^2.[1] The main theorems we seek to prove are (1) that H^2 is Archimedean, and (2) the strong version of MPP. With suitable adaptation of the methods used above in the recovery of E^2, this can be achieved with just regions-based axioms.

[1] The reqirement of Axiom 2.1 that all four sections of a gq under quadra-sectioning be mutually congruent, together with iteration of quadra-secting, can be shown to imply that, eventually, some quadra-sections will be similar to a larger gq.

6.1.2 *Core axioms and proof of Archimedean property*

We want, then, a variant of Axiom 2.1 that avoids similarity of non-congruents but that results, on iteration, in properly nested gq-sequences suitable for defining points.[2] At the same time, we want axioms providing for a potential covering of the whole hyperbolic space with (intuitively, connected) regions, say polygons, so that we can formulate and derive the Archimedean property (thereby proving that we really have such a covering). It is well known that, in fact, there are many tessellations of H^2, including ones by quadrilaterals.[3] Our strategy will be to formulate axioms that, in effect, allow us to derive the above-mentioned two main theorems from such a tessellation.

In addition to axioms of atomless mereology and axioms for either a logic of plurals or a weak set theory, we take both *Congruent* and *Adjacent* as primitive.[4] Some conditions of our axioms specific to H^2 pertain only to the quadrilaterals of the (4,5) tessellation (called the "order-5 square tiling"), and these we designate by 'sq' ("special quadrilaterals"), reserving 'gq' for arbitrary quadrilaterals. Next we define relations, also holding between two gq's, in ordinary terms intended to mean that a side of one of the gq's is equal in length to a side of the other, where we use the relative direction

[2] We want to be able to define 'point', not in order to recover the real number system over H^2, as the recovery from the one-dimensional interval structure already provided is sufficient, but rather for the sake of the geometry of H^2. We then readily obtain the "unity of space" and a means of introducing arbitrary "directions" from any given "point"-location, useful in deriving the Archimedean property. Also, we need 'point' and 'line' in order to state and derive the MPP.

[3] There are, in fact, infinitely many tessellations of H^2, even regular ones (i.e. by regular n-gons). Each is given a "Schläfli symbol", '(n, k)', where n is the number of sides of the n-gon, and k is the number of n-gons meeting at each vertex. In the case we've chosen, $n = 4$ and $k = 5$. Its dual, with $n = 5$ and $k = 4$, is also a regular tessellation. The (4,5) one is useful for introducing points, preserving much of what we did in the case of Euclidean 2-space, whereas the (5,4) tessellation preserves much of what was done regarding geodesics. As is evident from the pictures, in H^2 the corresponding "borders" of regular 4-gons and their extensions do not lie on a common geodesic (line). Rather lines correspond to bi-infinite sequences alternating back and forth between a border of a 4-gon and a segment diagonally bisecting the nearest contiguous but non-adjacent 4-gon (among the 5 at the relevant vertex of the 4-gon furnishing the border segment). (This can be deduced from the fact that 180° is a half-integral multiple, viz. $2\frac{1}{2}$, of 72°.) In contrast, with the (5,4) tessellation, the corresponding borders of sequentially adjacent 5-gons do trace out geodesics (as follows from the fact that the angles are all 90°).

For comparison with other 2-spaces of constant curvature: In E^2, $\frac{1}{n} + \frac{1}{k} = \frac{1}{2}$, determining just three regular tessellations (equilateral triangles, $k = 6$, squares, $k = 4$, and hexagons, $k = 3$). (The formula is based on partitioning n-gons into triangles and then considering the angle sum of triangles in the space in question.) On the sphere, S^2, there are three regular tessellations with equilateral triangles, viz (3,3), (3,4), and (3,5), and the famous icosahedral projection (soccer ball), (5,3), all satisfying $\frac{1}{n} + \frac{1}{k} > \frac{1}{2}$. In H^2, infinitely many satisfy the requirement, $\frac{1}{n} + \frac{1}{k} < \frac{1}{2}$.

[4] Adjacency, relating two gq's, is understood, informally, as "sharing an entire border exactly in common with". Even for the special quadrilaterals (sq's) we work with here, it cannot be defined as it was in the Euclidean case: in general, even for a tessellation of H^2 by sq's (explained further below), the fusion of one with an adjacent one is not itself a quadrilateral. In fact, although the corresponding borders of two adjacent sq's are exactly aligned (e.g. the top and bottom of an sq with those borders of its extension to the right), neither the top ones nor the bottom ones lie on a common geodesic. This is related to the fact that, unlike Euclidean tessellations via parallelograms, the regular tessellation of H^2 by sq's has five, not four, meeting at each vertex.

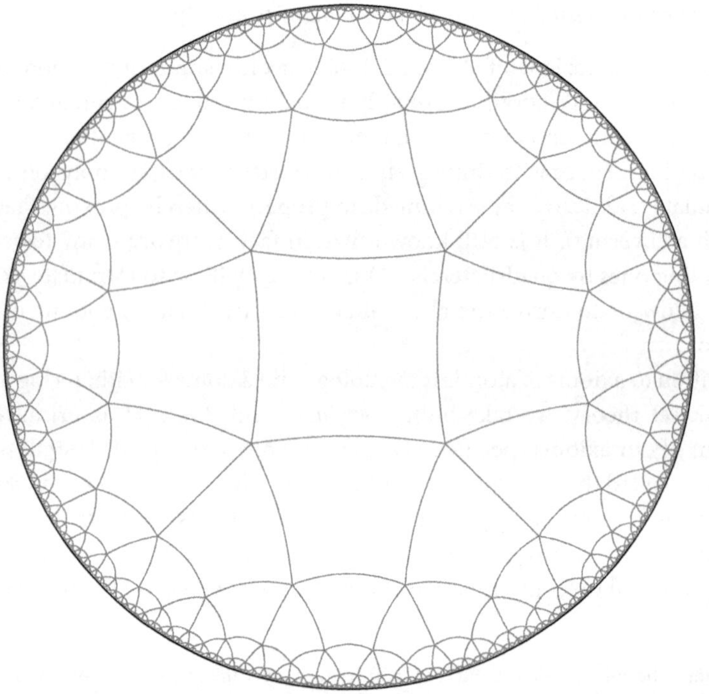

labels, 'L' for "left", 'R' for "right", 'T' for "top", and 'B' for "bottom". Letting 'D' and 'D′' range over these labels, the definition is the following scheme, where 'R' and 'S' range over gq's:

D-D'equiv(R, S) iff
$\exists S'[Cong(S, S')$ & S' is D of R & D' is D^- & R is D' of S' & $Adj(R, S')]$.

In words: "The D-side of R is the same length as the D'-side of S iff a congruent of S is adjacent to R and positioned to the D of R with its D′ side facing R in the direction D$^-$ opposite to D". (In effect, S' is the result of a suitable rotation and translation to the relevant position with respect to R.)

With these relations, we can readily say what it means to "bisect" a gq, "vertically" or "horizontally", expressing the effect of joining the "midpoints" of two opposite sides with a "line segment". Similarly for trisecting, indeed n-secting for positive integers n. Furthermore, by setting $S = R$, we can define "equilateral gq" (readily generalized to "equilateral polygon"). This will be useful when we come to introduce angles, below.

We also take over Axiom 2.1 from section 5.2, and the relative direction primitives and definitions and axioms governing them (replacing 'gq' with 'sq'), but not the clause about congruence (as above), and the clause that if s and t are continuous (and adjacent), their fusion is a gq, which can fail in H^2. Our modified Axiom 1a is the following:

Axiom H^2 1: *There are regions R such that R has a "quadra-section", resulting from bisecting R horizontally and vertically, that is,* $R = S_1 + S_2 + S_3 + S_4$, *with all the*

S_i discrete from one another, and such that for each S_i, there are exactly two S_j such that $Adj(S_i, S_j)$. Letting S_1 be "Top-Left", and proceeding clockwise with the labeling, T-T-equiv(S_1, S_2), R-R-equiv(S_2, S_3), B-B-equiv(S_3,S_4), and L-L-equiv(S_4,S_1). R has a "quadra-extension", that is, there are exactly four R_i, pairwise discrete and discrete from R, with $Cong(R_i, R)$ and $Adj(R_i, R)$ for $i = 1, \ldots, 4$.

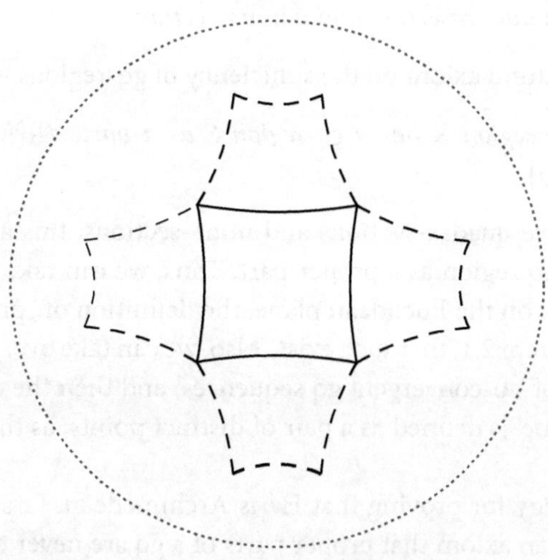

Finally, R has a "nona-section", that is, $R = \sum_{i=1}^{9} S_i$, with $S_i|S_j$, $i \neq j$, and with S_1, S_2, and S_3 mutually T-T- equivalent, S_3, S_4, and S_5 R–R -equivalent, S_5, S_6, and S_7 B–B-equivalent, and S_7, S_8, and S_1 L–L-equivalent (with S_9 central). Each extension R_i of R determined by quadra-extension is continuous with exactly three sections S_i of the nona-section; each of exactly four of the S_i is continuous with exactly two of the R_j, exactly one is continuous with no R_j, and the four others of the S_i are continuous with just one of the R_j.

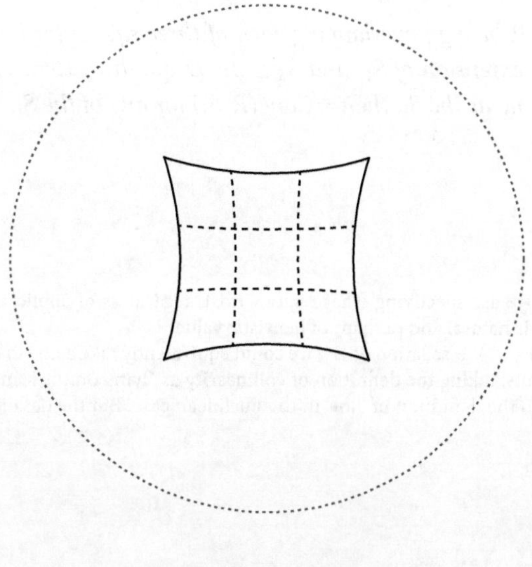

Our next axiom stipulates that sectioning and extending are inherited by congruents, and that the sectioning operations are indefinitely iterable:[5]

Axiom H^2 2: *Any region congruent to one with a quadra-section, a nona-section, and quadra-extension also has all of these. Any region congruent to one with a quadra-section or nona-section has both. Any region obtained by quadra-sectioning or nona-sectioning has both a quadrasection and a nona-section.*

One further structural axiom on the sufficiency of gq-regions will prove useful.

Axiom H^2 3: *Every region R has a gq-region S as a part:* $\forall R[R \text{ a region} \rightarrow \exists S(S \text{ a gq-region} \wedge S \leq R)]$.

Since gq regions have quadra-sections and nona-sections, this also guarantees that every region has a gq region as a proper part. Thus, we can take over from the core system of Chapter 5, on the Euclidean plane, the definition of "proper nesting" of gq sequences and Theorem 2.1, that such exist. Also, we can take over from the Euclidean case the definition of 'co-convergent gq sequences', and then the definition of 'point'. For H^2, however, 'line' is defined as a pair of distinct points, as this is valid for H^2 as well as for E^2.[6]

Here is our strategy for proving that H^2 is Archimedean. First, we'll simplify the task by adopting as an axiom that proper parts of a gq are never congruent to it; and also that a gq, R, overlapping all of three consecutive sq's, S_1, S_2, and S_3, where S_3 is the D-quadra-extension of S_2, and S_2 is the D-quadra-extension of S_1, cannot be congruent to any of the S_i. Then we'll formulate a modified version of Translation that applies to sq's and to bounded fusions thereof. We'll also need some method of treating sq's of arbitrary orientations, so that our proof generalizes to show, in effect, that in H^2, every region *in any direction* from a starting point (sq) can be reached in a finite number of steps.

Axiom H^2 4a: *Let S be a region and let R be a gq: then if* $S < R, \neg Cong(S, R)$.

Axiom $H^2$4b: *Let R be a gq overlapping each of three sq's,* S_i *for* $i = 1, 2, 3$, *such that* S_2 *is the D-quadra-extension of* S_1, *and* S_3 *is the D-quadra-extension of* S_2, *and where R is oriented the same as the* S_i; *then* $\neg Cong(R, S_i)$ *for any of the* S_i.

[5] Strictly speaking, we are specifying what regions exist, regardless of application of operations. But speaking of operations is natural and perhaps of heuristic value.

[6] If we work with the (5,4) tessellation of H^2, we could equivalently take a line to be a maximal collinear set (or plurality) of points, taking the definition of collinearity as "lying on the same border of a common sq", thus staying close to the definition of 'line' in the Euclidean case. But the definition via pairs of points is simpler.

We also need a condition formalizing that gq's (hence also sq's) and strings of them under quadra-extension in a given direction are *connected regions*, lacking any gaps, speaking informally here. We can formalize a special case of this as follows. First, we define R_D to be the fusion of the minimal closure of a given sq, R_0 under quadra-extension in D. Our next axiom expresses a necessary (but far from sufficient) condition for R_D to be a "connected region".

Axiom $H^2$4c: *If part of a gq, R, lies to the D of R_D but also overlaps an sq R_k that is part of the minimal closure whose fusion is R_D, then it also overlaps any sq R_{k+j} for any j, where R_{k+j} is in that minimal closure.*

We need to make sense of arbitrary "directions" relative to a given sq, R. We present two methods, the first a modest compromise with our "honest toil" approach (which we fondly call "petty theft"), in that it leads us to an axiom involving the defined concepts 'point' and 'line'. The second, thoroughly "honest", is a fairly straightforward adaptation of the apparatus of angles used above in recovering two-dimensional Euclidean structure.

The first method goes as follows: Identify the "central point", c, of R, with [the equivalence class of] the convergent sequence $\sigma = \langle \sigma_i \rangle$ of gq's, starting with $\sigma_1 = R$, setting σ_{i+1} = the central gq under nona-section of σ_i. Then a direction from c can be identified with a "line" defined by (c, b), where b is an arbitrary "border-point" of R.[7]

Given this much, it suffices for our purposes now to stipulate that sq's exist in arbitrary orientations.

To this end, we identify "the orientation of sq R" with the "line" determined by the pair (c, b_r), where b_r is the "mid-point" of the right border of R.[8] Then we posit:

Axiom H^2 5: *Let R be an sq, with center c; then for any border point b of R, there is an sq, R', with center c, $Cong(R, R')$, and with orientation (c, b).*

[7] Where 'D' ranges over 'top', 'right', 'bottom', and 'left', we define *p is a D-border point of gq R* just in case p is determined by a properly nested sequence $\langle \rho_i \rangle$ of gq's part of R where each ρ_{i+1} is one of the D-most among the gq regions of the quadra- or nona-section of ρ_i.

As in previous chapters, it should be clear that such "points" do not themselves enter into mereological relations. Instead, they are part of a "superstructure" defined over regions of the space. Also, they will be "common" to adjacent gq's, i.e. determined by properly nested sequences, gq's of one sequence part of one of the adjacent regions, those of the other sequence part of the other adjacent region. In point-based geometry, such points belong to a common boundary of two regions, hence to their closures. But of course we do not have the distinction between "open" and "closed" at the level of regions. As in the Euclidean case, that distinction belongs to the pointy superstructure.

[8] Where sq R is a regular quadrilateral of the (4,5) tessellation of H^2, let $\langle R_i \rangle$ be the sequence of congruent sq's obtained by iterating right quadra-extension, starting with $R = R_0$.Then the midpoints of the left and right borders of the R_i all lie on a geodesic of H^2. This justifies our definition of "orientation of an sq".

We can identify "mid-points" of the four borders of a gq R as illustrated for say the "top" border thus: form a properly nested sequence $\langle \sigma_i \rangle$ by setting $\sigma_0 = R$, σ_1 = top-left quadra-section of R, and from there on setting σ_{i+2} = top-right quadra-section of σ_{i+1}.

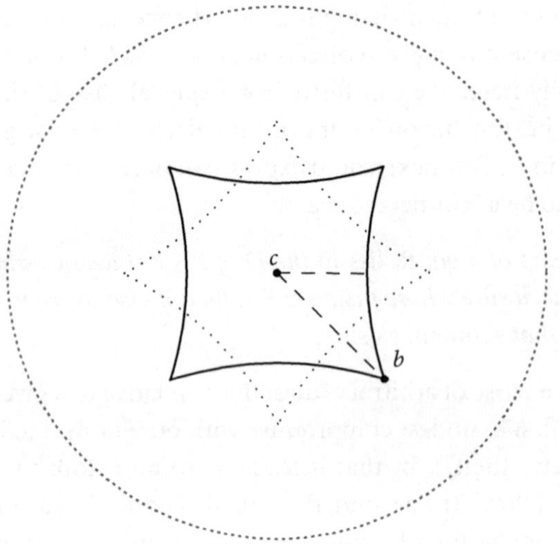

The purer, honest toil method proceeds as follows: We take over the primitives and axioms on angles of section 5.3, but with the following alterations. The norm function in terms of radians is dropped here in favor of degrees (or proportions of circles), insofar as reference to norms of angles is needed at all. The definition of circle is replaced with this:

C is a *circle* iff *C* is the fusion of (finitely many) sequentially adjacent angles, $\alpha_1, \ldots, \alpha_{n-1}, \alpha_n$, where all adjacencies are + or all are −, and where either $\alpha_n \circ \alpha_1$, or α_n and α_1 are adjacent (in the same sense as that of the sequence).

Next the honest toil method employs the rotation relations developed in section 4.3, simplifying Axiom 3.8 to:

Axiom H² 5′: *Let R be an sq with θ one of its angles, and let φ be any angle; then there is an sq, R′ such that Cong(R, R′) with θ′ an angle of R′ such that Cong(θ, θ′) and +(−)R(θ, θ′, φ),*

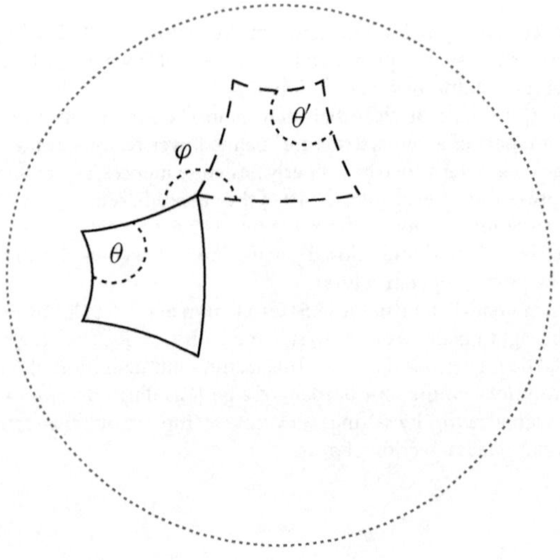

where the last formula refers to the rotation relation(s) defined above. The effect of either of these last two axioms is to endow our space H^2 with sq's oriented in arbitrary directions.

We now formulate a suitable translation axiom. Given an sq, *R*, use the variable '*D*' to range over the "directions" associated with the conventionally labeled borders of *R*, viz. "right", "bottom", "left", and "top". Then we import the definition of "*region s is entirely to the D of R*" from Chapter 5, on E^2 above. Further, let $\langle R_i \rangle$ be a (finite or countably infinite) sequence of sq's, starting with a given $R = R_0$, where R_{i+1} = the *D* quadra-extension of R_i. Now let *F* be the fusion of such a sequence. We call such *F* an "*sq* Σ*-region*". Then we define "*region s is entirely to the D of F*" just in case *s* is entirely to the *D* of each of the R_i of the sequence whose fusion is *F*. Next, we define "*F is a D-bounded sq* Σ*-region*" just in case there exists a region *s* entirely to the *D* of *F*. Here, then, is our version of Translation:

Axiom H^2 6 *(Translation): Let T be an sq or a D-bounded sq Σ-region; then there is a T' congruent to T and D of T such that the D^- quadra-extension T'' of T' overlaps T, $T'' \circ T$.*

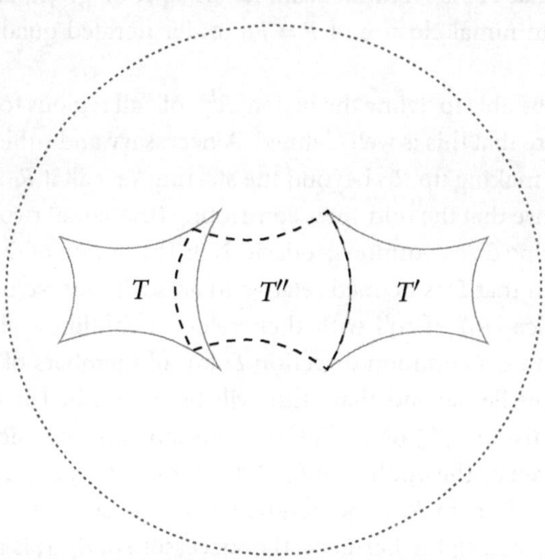

Thus, if *T* is D-bounded, we can get a copy T' of *T* to the *D* of *T* and close enough to *T* so that another copy, T'', adjacent to and to the D^- of T' overlaps *T*.

We haven't yet stated any axioms on "betweenness" governing lines, more specifically directions. But the following will suffice for deriving the Archimedean property:

Axiom H^2 7: *Let R be an sq and let R' be its D-quadra-extension. Then, if a region s is entirely to the D of R but is not entirely to the D of R', then $s \circ R'$.*

Note that a quadra-extension R' of an sq, *R*, shares an entire border with *R*. As in the development of E^2, we say that *R* and R' are "aligned". Thus, in effect, if *s* is entirely to the *D* of both *R* and R', where R' is also entirely to the *D* of *R* and adjacent to and aligned with it, then R' lies (entirely) between *R* and *s*.

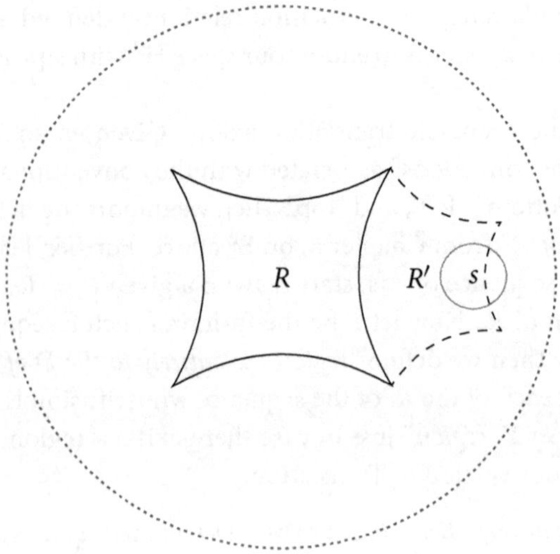

Now, to prove that H^2 is Archimedean, let an sq R be given, and let 'R_D' denote the fusion of the minimal closure of $R = R_0$ under iterated quadra-extension in D (relative to R).

We also want to be able to define the fusion, R^D, of "all regions to the D of sq R". But first we must be sure that this is well-defined. A necessary and sufficient condition for this is that the sq's making up R_D beyond the starting sq, call it R_0, define a common geodesic, in the sense that the mid-lines connecting (the same) two opposite sides are all line segments lying on a common geodesic. NB: The notion of D-quadra-extension of R makes sense in that D is defined relative to an sq R. But we need to make sense of a whole sequence $\langle R_j \rangle$ of sq's with their relevant midlines lying on a common geodesic, i.e., defining a common direction D for all members of the sequence and anything that might lie beyond that. This will be established if we can prove that the D-quadra-extension R_{j+1} of sq R_j has as midline an extension along the same geodesic as that given by the midline of R_j. But this follows by a symmetry argument:[9] a relevant midline of an sq R_j is perpendicular to the sides it joins (easily proved from congruent triangles); furthermore, the successor sq, R_{j+1}, is also the congruent reflection of R_j about the side common to the two sq's; and reflections are isometries. Now perpendicular lines get mapped into perpendicular lines by isometries, so the reflection that moved R_j into R_{j+1} maps the midline m of R_j into the midline m' of R_{j+1}. Since m is perpendicular to the common side of R_j and R_{j+1}, which is the line of reflection (producing R_{j+1} out of R_j), m and m' lie along the same geodesic, as was to be established. Thus we have proved:

Lemma: *Let $\langle R_j \rangle$ be a sequence of sq's obtained by the operation of quadra-extension in relative direction D. Then the midlines of these sq's all lie along a common H^2 geodesic.*

[9] Thanks to Victor Pambuccian for correspondence suggesting this argument.

Corollary: *Given an sq, R, the term 'R^D' standing for "the fusion of all regions to the D of sq R" is well-defined, indicating a common geodesic passing through all members of R^D.*

Theorem H^2 1: *Let an sq R be given. Then for any of the four principal directions, D,* relative to R, $R^D = R_D$.

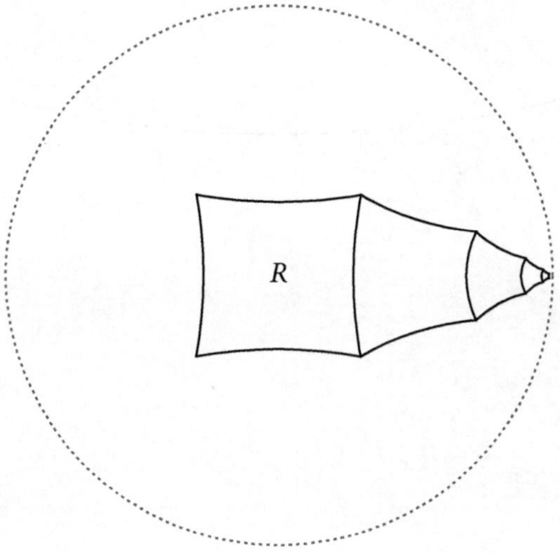

Proof. Clearly $R_D \leq R^D$. So suppose, for a contradiction, that $R_D \neq R^D$, i.e. that $R_D < R^D$. Then there exists a region $s \leq R^D$ such that $\neg(s \circ R_D)$. We claim that s lies entirely to the D of R_D. This is proved by induction on the index of the R_i of the sequence whose fusion is R_D: Since $s \leq R^D$, by definition it is entirely to the D of $R = R_0$. Suppose that s is entirely to the D of R_i, for a given $i > 0$. Then, since $\neg(s \circ R_{i+1})$, by Axiom 7, s is entirely to the D of R_{i+1} as well, establishing the claim. Therefore, R_D meets the conditions for being a D-bounded sq Σ-region. Therefore, Translation applies, and there must be a congruent R' of $R = R_0$ to the D of R_D close enough so that the D^- quadra-extension, R'', of R' overlaps R_D. Therefore we have that for some k, $R'' \circ R_k$, with R_k part of R_D. But then, R'' must, by Axiom H^2 4c, also overlap R_{k+1} and R_{k+2} (and indeed all the sq's of R_D that are D of R_{k+2}). Furthermore, by construction, R''—as the D^--quadra-extension of R'—is oriented the same as the sq's of R_D. Therefore $R'', R_k, R_{k+1}, R_{k+2}$ meet the hypothesis of Axiom $H^2$4b (with the R_i in the respective roles of the S_i), whence, by that axiom, $\neg Cong(R'', R_i)$, for $i = k, k+1, k+2$. But this contradicts that $Cong(R'', R_0)$, since, by the fact that Cong is an equivalence relation, that implies $Cong(R'', R_i)$, for all i. Thus, our hypothesis that $R_D \neq R^D$ is refuted, i.e., $R_D = R^D$. ■

But the orientation of our starting sq, *R*, is arbitrary, i.e. in virtue of Axiom 5 or 5′, there are congruent sq's in all orientations, and this theorem holds for all of them. This, then, completes the proof that H^2 is Archimedean.

6.1.3 *Deriving the MPP (strong form)*

Conveniently, it is well known how to do this, on the basis of an equivalent of MPP framed entirely in terms of regions.[10] The key concept is that of a "Saccheri quadrilateral", as usually described, a quadrilateral with a base, b, two equal-height

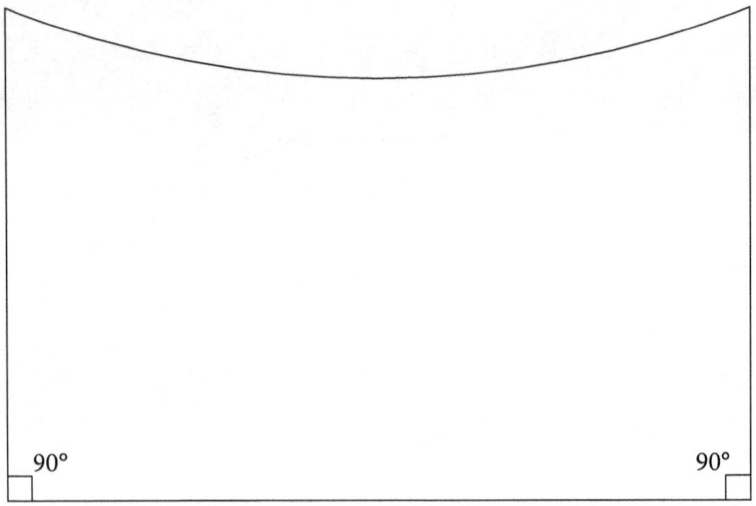

sides each at right angles to the base, and a line (H^2 geodesic) connecting the tops of the two sides. Then the new H^2 axiom simply states:

Axiom H^2 8: *The two top angles of a Saccheri quadrilateral are acute.*

In view of our apparatus of angles, this axiom is already expressible in our regions-based account. As usual, an angle is called '*acute*' iff it is $< 90°$ (or between 270° and 360°), and '*obtuse*' iff it is between 90° and 180° (or between 180° and 270°). In general, we assume that any real number between 0 and 360 measures a distinct angle. (This all makes sense in hyperbolic as well as Euclidean geometry.) Finally, to express "Saccheri quadrilateral", we use suitable instances of the (16) relations defined above to express "sides of equal height". A *Saccheri quadrilateral, S*, then, is a gq, with four angles, the two bottom ones of which are right angles and such that "the two arms are of equal height", i.e. *L–R-equiv (S,S).*

[10] For example, see [46].

The derivation of HPP from Axiom H^2 8 is elementary, making use of triangles and triangle–congruence relations, as well as properties of Saccheri quadrilaterals.[11] Triangles have already been introduced in the previous chapter. It is convenient, here, to build into the definition of 'triangle' that each of two triangles generated from a common gq are *D*-*D*′- equivalent with the other (where, for a partition along diagonal, say, UR-BL, *D* will be BR for the upper triangle, and *D*′ will be UL for the lower one, etc.).

Then define two triangles, T, T' to be *congruent* iff there is an order- or reverse-order preserving 1–1 correspondence φ between the three angles, α_i, $i = 1, 2, 3$, of T and those of T' such that α_i and $\varphi(\alpha_i)$ are congruent, for each i. (Note that this is correct in H^2, though not, of course, in E^2.) We then want to adopt an axiom corresponding to the standard criteria for triangle congruence, "SSS" and "SAS". To keep this within a language of regions, we can use our gq orientation labels plus a suitable label for the relevant diagonal to label appropriate "edge–equivalence" relations. For example, we can say that if triangles T, T' are R–R′-equivalent, B–B′-equivalent, and UL–U′L′-equivalent, then SSS is satisfied. Also sufficient would be R–L′-equivalent, B–T′-equivalent, and UL–B′R′-equivalent, and likewise for the other two cases, covering the two diagonal partitions of each of the two parent quadrilaterals. Similarly, we can express the sufficiency of SAS for triangle congruence. Thus, our final (abbreviated) axiom reads:

Axiom H^2 9: *Each of the cases of SSS and of SAS is sufficient for congruence of two triangles, T and T'.*

Now it is straightforward to prove HPP from Axiom H^2 8 on Saccheri quadrilaterals, as well as the theorems that the angle sums of triangles are always $< 180°$ and approach that value as triangles are made smaller and smaller, so that similar triangles are congruent, thereby incorporating the geometry of H^2 into our regions-based theory.

6.2 Spherical Geometry: Letting Bigons Be Bigons

Our aim in this section is to apply our methods to the case of spaces of constant positive curvature, in particular two-dimensional spherical geometry. The entire space, called S^2, is our intended structure. Our main task is to formulate axioms on regions (of S^2) which allow us to prove (1) that there are no infinitesimal regions; and (2) that there are *no* parallels to any given "line" through any "point" not on the given line. Of course, the latter requires that we define '*point*' and '*line*', so our axioms should guarantee the means of doing this. They should also imply, via such definitions, that the other Euclidean axioms hold except for the axiom that any two points determine a

[11] See, for example, [46], pp. 10–16.

unique line, the exceptions being antipodal points. Thus, we will want to define '*points p and q are antipodal*', as well.

6.2.1 *Axioms*

Our logical and mereological axioms are the same as those given for Euclidean and Hyperbolic spaces. In particular, we have the second-order strength to guarantee the existence of arbitrary (non-empty) fusions.

Our first extra-logical primitive (beyond part–whole of mereology) is the one-place predicate, '*Bigon*'. In ordinary, point-based terms, a *bigon*—also called a *lune*—is an equilateral two-sided polygon, with two equiangular vertices at antipodal points.[12]

The angle is stipulated to be $< 180^o$. (This can be stipulated in terms of regions: it can, for instance, be required that a bigon always be a proper part of a "hemisphere", as defined below.) All bigons of S^2 have "sides" of the same length, half the circumference of the space, and (by definition, in point-based terms) the two vertices of any bigon are *antipodal*. This will be expressed in terms of regions only below.

We also adopt $Cong(B, B')$, intuitively meaning that bigons B and B' are congruent. Below we will show how to define '*angle*'; then '*angle congruence*' will be defined as congruence of the bigons of which they are, respectively, vertex angles. Also, we adopt $Adj(B, B')$, that bigons B and B' are *adjacent*, meaning, informally, that they share an entire border in common. In terms of regions, there is no region between two adjacent bigons. Given a bigon, B, there is a unique "diametrical", "antipodal", or "complementary" bigon, B', in the opposite hemisphere, with $Cong(B, B')$. In terms of points and lines, each of the two half-lines of B is extended, respectively, by one of the half-lines of B' to form a complete line or geodesic ("great circle").

[12] Of course, bigons cannot exist in Euclidean or hyperbolic spaces, depending as they do on the positive curvature of the space.

It is assumed that a bigon always comes with an orientation, i.e. a 'north' and a 'south' relative to B. In these terms, it also will make sense to speak of bigons B' adjacent to B as "*to the west of B*" (relative to B) or "*to the east of B*". (Our colleagues from "down under" are of course free to choose the opposite convention.)

Similarly it is convenient to speak of *triangles*, also related by *congruence* and *adjacency*. Likewise, we speak of quadrilaterals ("*gq's*"). We also introduce relations of "*edge-equivalence*" relating two triangles, two gq's, or a gq and a triangle, intuitively meaning that they are of equal length along the edges being compared, e.g. "*WE(R,S)*", to say that the west edge of R is equal in length to the east edge of S, etc., similarly for "*LR(R,S)*", where convenient, to express equality of length of the "left side" of R with the "right side" of S, etc. In particular, a triangle T with a base and "left" and "right" sides is "*isosceles*" iff $LR(T, T)$.[13]

It will also prove convenient to introduce ring-shaped regions or *annuli*, $Ann(A)$, along with the interior and exterior regions defined by a given annulus: $Int(I, A)$, "*I is the interior of A*", and $Ext(E, A)$, "*E is the exterior of A*", so that $A + I + E = U$, the mereological universe, $= S^2$. (Where necessary, we write I_A and E_A to indicate the interior and exterior determined by annulus A.)

[13] Although not needed for the two significant theorems derived below, a fuller exposition would involve some obvious axioms governing regularities among these relations, e.g. that some are equivalence relations (viz. when the same sides are being compared), that all are reflexive and "symmetric" (in the sense that e.g. $LR(R,S) \leftrightarrow RL(S,R)$ etc.), and that all "transitivities" hold (i.e. where the correspondence between figure and side is retained throughout the formula).

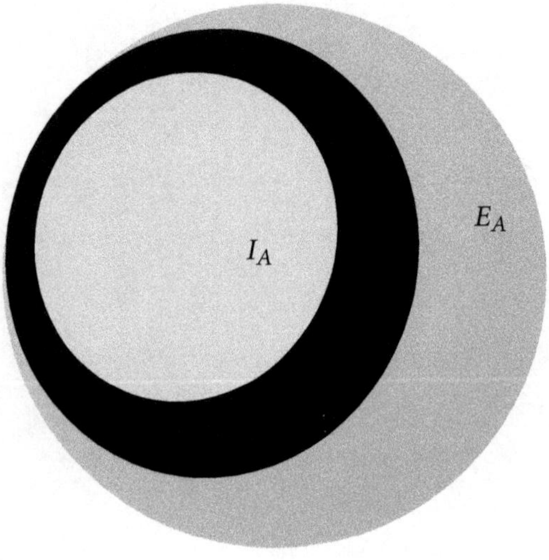

Our first axiom stipulates the existence of a convenient covering of the space by four mutually congruent bigons:

Axiom S^2 1: *There are four mutually congruent bigons, $B_1 - B_4$, each adjacent to exactly two others, such that their fusion $= U = S^2$.*[14]

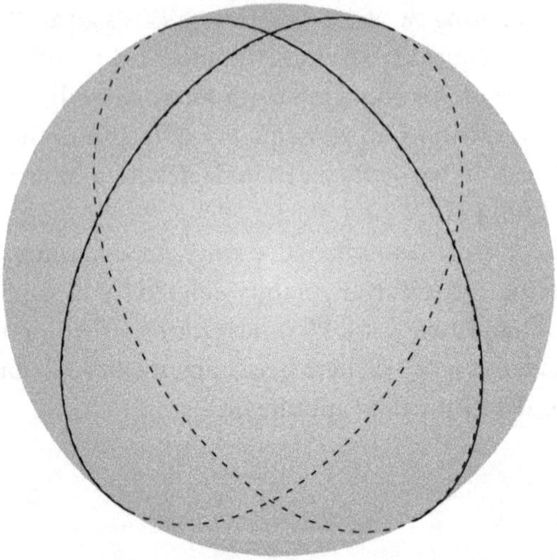

Once 'angle' and 'angle-congruence' have been introduced, the angles of these particular bigons will be recognized as all right angles. And it will follow that the two vertex angles of any bigon are always angle-congruent to each other. Collectively these

[14] To be precise, we should confine the designator 'S^2' to our background metalanguage, using 'U' for the universal object, available by explicit definition in the object language.

bigons *cover* the space, or form a *covering*: that is, every region r is part of the fusion of the bigons of the covering. Of course, there are infinitely many coverings of the space (S^2) by mutually congruent bigons, even meeting at the same vertices (as points, which can be introduced as described below). The fusion of any adjacent pair of these bigons constitutes a *hemisphere*, with the opposite hemisphere formed by the other pair of bigons.

Furthermore, it will be useful to decompose bigons as prescribed by the next three axioms.

Axiom S^2 2a: *Any bigon B has a vertical bisection, viz. there are unique bigons,* B_x, B_y *such that* $B_x < B$ and $B_y < B$, $Cong(B_x, B_y)$, $Adj(B_x, B_y)$, *and* $B_x + B_y = B$.

Once angles have been introduced below, it will follow that the vertex angles of B_x and B_y are a bisection of the respective vertex angles of B.

Axiom S^2 2b: *Any bigon B has a lateral bisection, viz. there are unique isosceles triangles,* T_1, T_2 such that $Cong(T_1, T_2)$, $Adj(T_1, T_2)$, *with* $B = T_1 + T_2$.

Axiom S^2 2c: *Any triangle T has a lateral bisection into a lesser triangle* T_1 *and a gq Q such that* $T = T_1 + Q$ *and satisfying (say) left–left and right–right edge equivalences, i.e.* $LL(T_1, Q)$ *and* $RR(T_1, Q)$; *and any gq Q has a lateral bisection into two gq's,* Q_1, Q_2, *with* $Q = Q_1 + Q_2$, *satisfying those edge equivalences.*

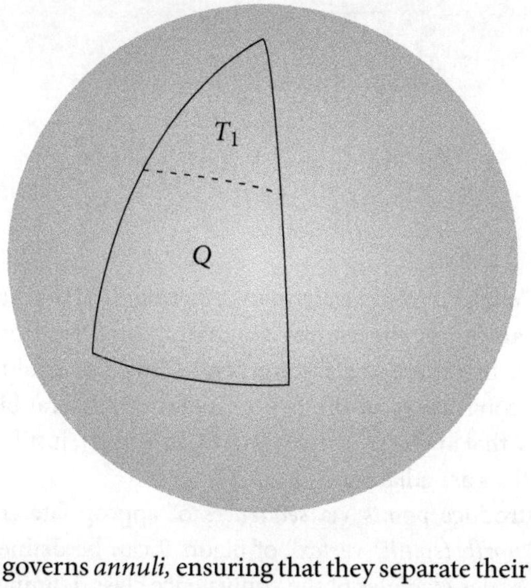

Our next axiom governs *annuli*, ensuring that they separate their interior from their exterior regions. To this end it is convenient to define the minimal closure, $Cl(B)$, of a bigon B under the operations of angle and lateral bisections. (Any region in such a minimal closure is suitably connected.)

Axiom S^2 3a: *Given any annulus A with interior I and exterior E and any bigon B and region* $R \leq Cl(B)$, *if* $R \circ I$ *and* $R \circ E$, *then* $R \circ A$. Furthermore, $A + I_A + E_A = U$, the universal object (i.e. satisfying $\forall r(r \leq U)$).

The next axiom distinguishes the interior from the exterior of an annulus.

Axiom S^2 3b: *Given a bigon B and annulus* $A < B$, $I_A < B$, whereas $E_A \nless B$.

Now we can introduce '*angle*' by definition. Specifically, we define angles as *vertex angles* or *polar angles* of a given bigon B. We define these two angles, "north vertex angle of B" and "south vertex angle of B", respectively, as equivalence classes of isosceles triangles obtained by first laterally bisecting bigon B, and then successively laterally bisecting the north (south) triangle obtained by the previous lateral bisection, *ad infinitum*. That is, we define the triangles $T_j^{N_B(S_B)}$ for $j = 1, 2, 3, \ldots$ via $T_1^{N_B(S_B)}$ = the north (south) triangle of the lateral bisection of B, and $T_{j+1}^{N_B(S_B)}$ = the north (south) triangle of the lateral bisection of $T_j^{N_B(S_B)}$.

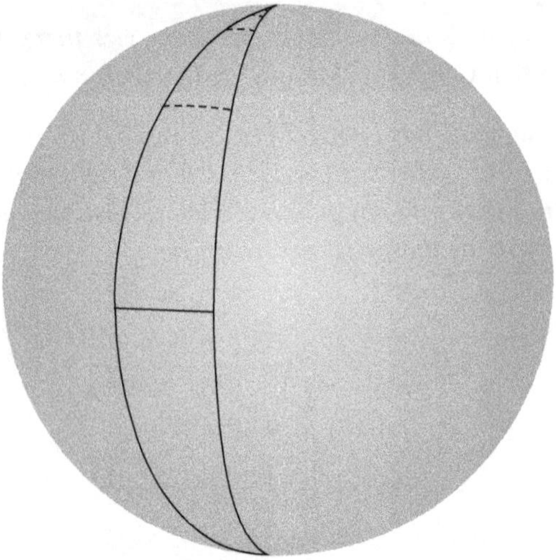

We then define "*angles* α, β *are congruent*", $AngCong(\alpha, \beta)$ iff the bigons of which α and β are vertex angles are themselves congruent. It follows from this that the two vertex angles of any bigon are angle-congruent. (This also could be deduced from Axiom 2b, via the congruence of the two triangles of a lateral bisection of a given bigon). Also, we say that angles α, β are *adjacent*, $Adj(\alpha, \beta)$, just in case the bigons of which they are vertices are adjacent.

We now can introduce *points* via *sequences* of appropriate convergent regions. In particular, the "*north (south) vertex*" of bigon B can be defined as the sequence $\langle T_j^{N_B(S_B)} \rangle$ of the isosceles triangles of the equivalence class defining the north (south) vertex angle of B, as above. (In this manner, no confusion of vertex angles and vertices as points arises.)

Now in order to introduce lines—as pairs of non-antipodal points—we will need to define (ternary) *collinearity* among points as a criterion for which pairs of (non-antipodal) points determine the same line. This will be accomplished below.

So far, our axioms do not guarantee that there are bigons at arbitrary locations and with arbitrary orientations, or even bigons at a dense set of locations and with a dense set of orientations. Given our definitions, our next axiom provides for the latter, given sufficient iteration of vertical and lateral bisections.

Axiom S^2 4: *Any triangle T has three angles, α, β, γ, and any quadrilateral Q has four angles; one of the angles, say α, of a triangle T is the bigon vertex angle to which T belongs; then β is adjacent to one of the angles of Q, and γ is adjacent to another.*

To state our next axiom, we need to distinguish, given a bigon, *B*, angles that "abut a border" of *B* from those that do not, the latter of which we may dub "properly interior" to *B*. Here the notion of *annulus* proves useful:

Given a bigon, *B* and angle $\alpha = \{T_j^\alpha\}$, we say α *is properly interior to B* iff there is an annulus, $A < B$, and such that $T_j^\alpha < I_A$, for some *j*. (This, of course, implies that $T_k^\alpha < I_A$, for any $k > j$.) Similarly, for a hemisphere *H* in place of *B*.

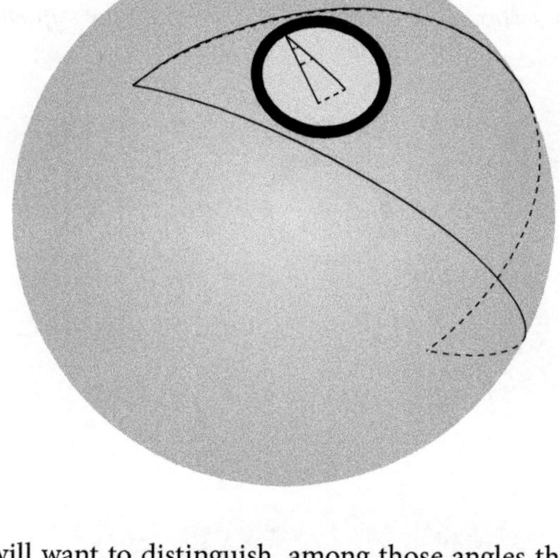

Moreover, we will want to distinguish, among those angles that are not properly interior to *B*, although still part of *B*, those that are "at the west of *B*" from those that are "at the east of *B*". In the familiar terms of points and lines, that is, we will want to distinguish the angles (as parts of *B*)[15] whose vertices lie on the west boundary of *B* from those whose vertices lie on the east boundary. One way of accomplishing this uses annuli (to play the role of open neighborhoods of point–set topology): For angles α that are part of bigon *B*, we define "*α is not properly interior at the west (east) of B*" to mean that α is not properly interior to *B* and for every sufficiently small triangle

[15] We say of an angle $\alpha = \{T_j\}$ and a bigon *B* that $\alpha < B$ just in case for some *j*, $T_j < B$. It follows then that for every $k > j$, $T_k < B$.

$T_j \in \alpha$, every annulus A with $T_j < I_A$ overlaps a bigon B' adjacent to and to the west (east) of B.

This now allows us to define *collinearity* as a relation among triples of points: Let $p_\alpha, p_\beta, p_\gamma$ be three pairwise non-antipodal points, as vertices of the angles α, β, γ that are vertex angles of bigons $B_\alpha, B_\beta, B_\gamma$, respectively. Without loss, we may assume further that there is a covering defining a hemisphere H such that all the angles α, β, γ lie in the same hemisphere. (For any outlier, substitute its antipode.) Then we define "*α, β, γ are collinear*", $Col(\alpha, \beta, \gamma)$, iff there is a bigon B such that α, β, γ are all either not properly interior at the west of B or not properly interior at the east of B. Then we say that pairs of points $\langle p_1, q_1\rangle, \langle p_2, q_2\rangle$ *define the same line l* just in case all triples among them are collinear.

Further useful definitions are the following:

Bigons B_1, and B_2 are *vertex (or polar) equivalent* or *co-polar* iff they derive from the same covering by vertical subdivision and fusion.

Given a bigon B (or a hemisphere H), and given another bigon B', we say that *a vertex angle β' of B' lies in B (H)* iff for some j, $T_j \in \beta'$ and $T_j \leq B$ (H). In that case, we also say that *the vertex of B' (as the point defined by the sequence $\langle T_j\rangle$ of triangles in β') lies in B (H).*

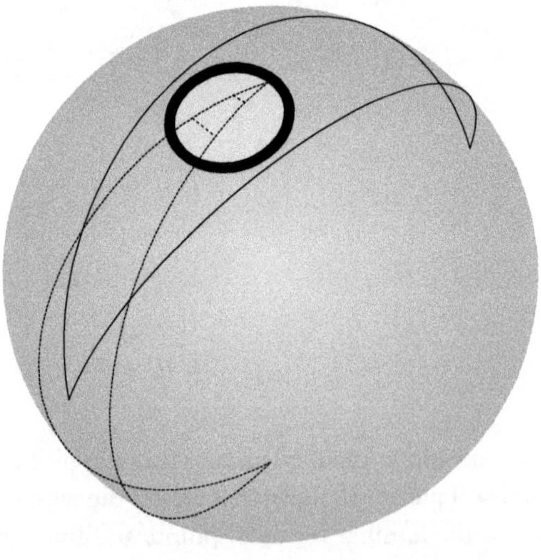

Axiom S^2 5a: *Given a hemisphere $H = B_1 + B_2$, the only bigons $B' < H$ are those whose vertex angles are not properly interior to H.*

Axiom S^2 5b: *Given a hemisphere, H, and a bigon, B, at most one vertex of B is properly interior to H; the other vertex then lies in the hemisphere H'* antipodal to H.

In the standard language of points and lines (geodesics), a bigon $B' < H$ just in case B' is either co-polar with the B_i constituting H or it has a vertex lying on the geodesic defining the boundary of H. For a familiar example illustrating 5a and b, given a point in the Northern hemisphere, its antipodal point must lie in the Southern, and vice versa. In such a situation, we say that any geodesic connecting points in opposite hemispheres *meets* the geodesic boundary of the hemispheres, i.e. being determined by points lying in opposite hemispheres is criterial for "meeting the boundary".[16]

For our final axiom, let $\{B_i\}$ be a finite covering of S^2 by bigons, and let B^* be the minimal closure of $\{B_i\}$ under vertical and lateral bisectionings of any of the B_i. The axiom then states:

Axiom S^2 6: *Every region r contains a region b of B^** as a part: $\forall r \exists b[b \in B^* \wedge b \leq r]$.

6.2.2 *Theorems*

A region is defined to be *infinitesimal relative to* S^2 $(= U)$ just in case it is too small to contain as a part any region in any B^* based on any finite covering $\{B_j\}$ of S^2. Then it is immediate from Axiom 6 that

Theorem S^2 1: *Relative to S^2, no region is infinitesimal.*

Now we can establish our principal geometric goal of this section, the absence of parallels, utilizing the defined predicates, 'point' and 'line', as defined above.

We need one more definition: Let p_α be a point defined by a sequence $\langle T_j \rangle$ of triangles of a vertex angle α of a bigon B. Then, for any given bigon B' or hemisphere H, we say p_α *is properly interior to* B', respectively H, just in case α is properly interior to B', respectively H.

Theorem S^2 2: *Given a line l and a point p not on l, every line l′ through p meets l.*

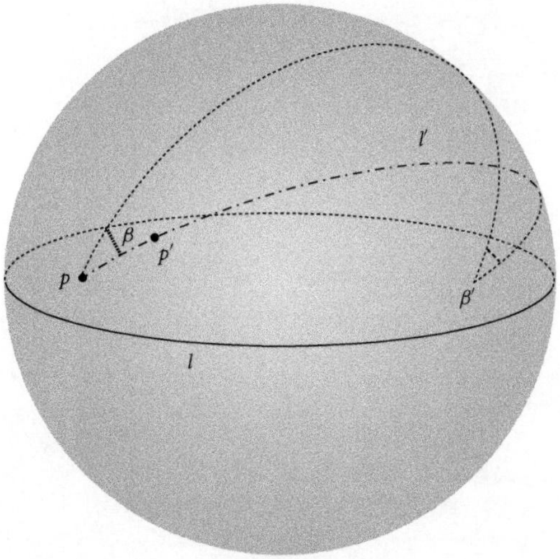

[16] In a point-based analytic treatment, of course, "meeting the boundary" would be defined as sharing a common point with it, and the fact would be proved as a consequence of the intermediate value theorem.

Proof. Let l and p be given with p not on l. Let H_1 and H_2 be the hemispheres determined by l, and suppose—without loss—that p lies in H_1. By hypothesis, p is properly interior to H_1, as it does not lie on l. Now let p' be any point distinct from p and not antipodal to it. If p' lies in H_2, then by definition, the line l' determined by $\langle p, p' \rangle$ already meets l and there is nothing to prove. If p' lies in H_1, then we argue as follows: Without loss we may assume that p is at the vertex of a bigon B one of whose "sides" is l', determined by $\langle p, p' \rangle$. Since the vertex angle, call it β, of B at p is properly interior to H_1, we have by Axiom 5a that $B \nleq H_1$. Let β' be antipodal to β. Then, by Axiom 5b, the angle β' (or associated vertex of B) antipodal to p must lie in H_2. Then by construction, l' meets the boundary of H_1 and H_2, which is l. ■

7

The Matter of Points

7.1 Introduction

We have seen that our various semi-Aristotelian frameworks, with atomless mereology, and further machinery of higher-order logic (or the logic of plurals, or some set theory), yields workable theories of structures central to classical analysis, Euclidean geometry of n-dimensions, and non-Euclidean geometries of two-dimensional spaces of constant curvature. Furthermore, these theories enable recovery, indeed full reductions, of their point-based counterparts, which emerge as meta-theoretic superstructures, available for further construction, development, and applications.

As indicated, the recoveries and reductions are reversible: one can start with a standard, point-based theory and proceed to introduce suitable "regions" as certain point-sets, explicitly define the relevant relations among these "regions", and derive, as theorems, translations of the axioms of our regions-based theories. The upshot is full equivalence of the two frameworks, at least for the spaces examined thus far. Neither points nor regions can be claimed indispensable for the mathematics of these spaces.

Two closely related questions thus arise. First, can regions-based versions of further point-based mathematical theories be developed, and, if not, what conclusions would follow from such limitations? Second, what sense can be made of the question of whether spatial or space–time points are real? This second question takes us to metaphysics.

One can wax Carnapian (e.g. [30]) and argue that the only legitimate question to ask is which of the frameworks for space or space–time is more suited to this or that theoretical purpose. In effect, that is our conclusion in this chapter.

We begin with a brief recap of the connections between our semi-Aristotelian, regions-based theories and their punctiform counterparts, illustrating, one final time, one way to define "points" in our frameworks, and also how to construct models for our regions-based theories in standard Dedekind–Cantor spaces. Then we turn to the first of the questions listed just above, whether it is possible to extend our regions-based accounts to cover some more sophisticated topics. Without any claim of comprehensiveness, we consider two areas of modern mathematics that appear to raise challenges to regions-based treatments, namely the theory of measure, specifically Lebesgue measure, and differential geometry, of special importance in connection with space–time physics.

Then we turn to more metaphysical matters. Contemporary analytic metaphysicians have occupied themselves with the nature (or the possible nature) of space and time (or space–time), and with the relationship between physical objects and the regions of space or space–time they (exactly) occupy. Some of the issues concern the boundaries of objects, and notions like contact; others concern matters of grounding or fundamentality. Much, but not all, of the metaphysical literature just assumes that the underlying space or space–time is punctiform, following the standard Dedekind–Cantor picture.[1] That is, many writers assume that space or space–time is composed of points—it is structured like $\mathbb{R}^3$ or $\mathbb{R}^4$. Physical objects thus occupy regions that are nothing but sets of points. Some of those authors get tangled up over what to say about point-sized regions, or even point-sized objects. We give a somewhat biased overview of a portion of this literature, arguing that many of the issues are much easier to negotiate if we assume a regions-based space or space–time.

We conclude this chapter with a brief treatment of the question of whether space or space–time really is punctiform. One might argue, against the Carnapian, that if one or the other of the accounts of space or space–time is so much more convenient than the other, for whatever purpose may be at hand, then perhaps space or space–time really is the way the account says it is. Suppose, for example, that it is much easier to think metaphysically about physical objects, and their boundaries and contact, in a regions-based framework. Does that give us a reason to think that points are not real? On the other hand, if it turns out that it is so much more convenient—either in everyday discourse of objects, boundaries, and contact, or, more likely, in science itself—to think of space-time as composed of points, and, for example, to think in terms of the values of certain fields at points, then perhaps we should think of points as real, and not as mere mathematical constructions on a more fundamental gunky space. We resist both conclusions, at least tentatively. In light of the tight connections between the (semi-Aristotelian) regions-based theories and their punctiform counterparts, the dispute over whether space or space–time is punctiform or gunky is merely verbal.

7.2 Connections

Begin with a standard, Dedekind–Cantor space, or just the real numbers. We noted above that the regular open sets constitute a model of our regions-based theories. In particular, "regions" are regular open sets of points (or real numbers, in the one-dimensional case), "parthood" is just the non-empty subset relation, and the other primitives of our theories are readily defined. A second model of each of our theories

[1] There are some more or less dismissive references to point-free spaces, often with reference to Tarski [119] or Whitehead (e.g. [128]; [129], part III; or [131], part IV, ch. 2). Some authors do take the possibility of gunky space (or space–time) seriously.

is obtained by taking the "regions" to be regular closed sets of points (or numbers), again with "parthood" being the non-empty subset relation.

Now shift gears, and think in terms of one of our semi-Aristotelian theories. In the treatments above, we gave several different ways to introduce "points", as meta-theoretic constructions—a kind of Whiteheadian "extensive abstraction". For the sake of readability, we briefly repeat one of those here; for the sake of simplicity, we stick to one–dimension, although the treatment readily extends to the more sophisticated theories.

Define a non-empty set S of intervals to be *convergent* if[2]

(i) if $i \in S$ and $j \in S$, then either $i \leq j$ or $j \leq i$, so the intervals in S are linearly ordered by parthood,
(ii) if $i \in S$, $j \in S$, and $i \neq j$, then neither is right-end or left-end-equivalent to each other, and
(iii) there is no region that is part of every member of S.

Let S and S' be convergent sets of intervals. Say that $S \sim S'$ if for every $i \in S$ and $i' \in S'$, i overlaps i'. It is straightforward to verify that this is an equivalence relation.

Define a *"point"* to be an equivalence class on this relation. Again, we identify each such "point" with a representative member of its equivalence class. The "points", so defined, are indeed a Dedekind–Cantor continuum, a model of standard real analysis.

The connection between the regions-based theories and their Dedekind–Cantor counterparts is especially tight. Work, for example, in the one-dimensional gunky theory of Chapter 2. Let p be a "point" as defined above, and let m be a region. We can say what it is for p to be *interior* to m: this is just for some interval in p to be a proper part of m. Similarly, we can say what it is for p to be *exterior* to m: that is for some interval in p to be discrete from m. And we can say what it is for p to be a *boundary* of m: that is just for p to be neither interior nor exterior to m, or, in other words, for every interval in p to overlap m and for no interval in p to be a part of m.

Now put the straightforward topology on the "points", so defined: a set S of "points" is *basic open* if there is an interval i such that S is the set of points interior to i. A set is open just in case it is a union of basic open sets.[3] We can then show that a set S of "points" is regular open in this topology if and only if there is a region r such that a "point" p is in S if and only if p is interior to r. And a set T of "points" is regular closed in this topology if and only if there is a region r such that a "point" p is in T if and only if p is either interior to r or on the boundary of r.

A formal theory A is said to be *interpretable* in a formal theory A' just in case there is an effective function f from the sentences of A to the sentences of A' such that if Φ is

[2] In two dimensions, we would define a non-empty set of generalized quadrilaterals to be convergent if its members are linearly ordered by parthood and no distinct members share a "boundary", and if there is no region that is part of every member of the set.

[3] In the two-dimensional case, we'd start with the sets of "points" that are interior to a generalized quadrilateral.

an axiom of A, then $f(\Phi)$ is a theorem of A'. It is sometimes required that the function f preserve logical form (so conjunctions go to conjunctions, universal quantifications go to universal quantifications, etc.). The theories are *mutually* interpretable if they are interpretable in each other. Finally, two theories A, A' are *definitionally equivalent* (also called *synonymous*) just in case there is a function f that interprets A in A' and a function g that interprets A' in A (so that the theories are mutually interpretable) such that for any sentence Φ of A, the sentence $gf(\Phi)$ is equivalent to Φ in A; and for any sentence Ψ of A', $fg(\Psi)$ is equivalent to Ψ in A. Equivalently, two theories are definitionally equivalent if there is a single theory that is a definitional extension of both. This is a rather strong articulation of the sense in which two formal theories are "notational variants" of each other (see [35] or [126]).

In light of the connections articulated already, it is straightforward to verify that each of our semi-Aristotelian, regions-based theories is definitionlly equivalent to its punctiform, Dedekind–Cantor counterpart. Thus, in each case, the two *mathematical* theories are about as close to each other as can be. Intuitively, they describe the same structure (see [106], 241-2). Our concern now is with the theories as part of a more general scientific or metaphysical enterprise.

7.3 Mathematical and Scientific Limitations

7.3.1 *Measure theory*

It might seem, at first blush, that a regions-based continuum is especially well suited to measure theory, as one might expect there to be a close link between regions and measurable sets of points. Consider, for instance, Cantor's famous middle-third construction of an uncountable point–set of measure 0 (demonstrating the vast gulf between measure and cardinality of point–sets). Start, for example, with the open interval $(0, 1)$. Remove the middle third. Then remove the middle third of each of the two remaining intervals. Then remove the middle third of each of the remaining intervals, and continue. That uncountably many points remain (in the limit) follows from the ternary representation (in terms of '0', '1', '2') of the points of the unit interval, since exactly the points removed have a '1' somewhere in their ternary expansions, and uncountably many points have just '0' and '2' occurring in theirs. The latter make up the "dust" remaining after all the middle third intervals have been removed. And since, by countable additivity, the sum of the (Lebesgue) measures of the removed intervals is the measure of the unit, i.e. 1, the dust has measure 0. That's the standard punctiform description.

Essentially the same construction can be carried out in our regions-based theory, but the results are somewhat different. In light of the axiom that every region has an interval as a proper part, the theory simply refuses to recognize the existence of the Cantor dust, as a region. In terms of regions, precisely *nothing* remains after all the steps of removing middle thirds. Equivalently, suppose we take the fusion of the middle third of a given interval, the middle thirds of the two remaining intervals, the middle thirds of the intervals that remain after that, etc. The result is just the original interval.

So far so good, one may say. The Cantor set, and its relative complement—the Cantor dust—are pathological. One might think it a plus that the regions-based theories do not recognize them, at least not as regions.[4] But, unfortunately, there are similar constructions in which the residual dust has *positive* Lebesgue measure. Frank Arntzenius [5] carries out a construction based on removing, at any stage, the middle *quarter* of all the remaining intervals (see also [6] and Jeffrey Russell [101]). Start with the unit interval $(0, 1)$. Remove the middle quarter. Then remove the middle quarter of the remaining intervals, etc. In the limit, the measures of the removed intervals add up to $\frac{1}{2}$; so the dust has measure $\frac{1}{2}$. In the regions-based theory, the result is the same as in the original Cantor construction: since the dust contains no interval as a part, it does not exist. In other words, if we take the fusion of the middle quarter of a given interval, the middle quarter of the remaining intervals, etc., the result is the interval with which we began. But the intuitive measures of the fused intervals (all of which are pairwise discrete) add up to only $\frac{1}{2}$ of the measure of the original interval. By failing to recognize dusty regions, one confronts violations of countable additivity.

What can the regions theorist say about this? One of three things: (1) one could take back the axiom that every region has an interval as a proper part;[5] (2) one could seek to live with just finitely additive measures; or (3) one could concede that the theory of Lebesgue measure is essentially pointillist, and simply continue using it as part of the superstructure, defined over the regions-based continuum.

The trouble with option (1) is that it is arguably an essential feature of a "gunky" line that any part of it be gunky as well, and so be resolvable into intervals. In a nutshell, Cantor "dust" is conceptually incompatible with "gunk", or so we suggest. As for option (2), while countable additivity is used in physics (e.g. in connection with probability), it is often possible to resort to a "coarse graining", whereby finite additivity suffices. But even if that is not always satisfactory, there is the following consideration. As already demonstrated, the Dedekind–Cantor theory of the punctiform continuum is fully reducible to the regions-based theory. From a logical point of view, measure theory requires no axioms beyond the Dedekind–Cantor theory, with the possible need to recognize some more sets of point-sets, which would at most be a "friendly amendment" to a regions-based theory at the level of superstructure.

In short, the proponent of a regions-based theory can legitimately "have it both ways", adhering throughout to an ontology of regions, yet theorizing with points, point-sets, and point-set-sets, as needed for applications. There is nothing in a

[4] Of course, the Cantor sets can be constructed in the punctiform superstructure imposed on the gunky space.

[5] This is Arntzenius's preferred solution. It is obtained from a Dedekind–Cantor space by "identifying" point-sets when their differences have measure zero.

regions framework that implies that defined superstructural items may not be genuinely useful or even "required", say in physical applications of mathematics.

It is one thing to say that the point-concept may be unavoidable in certain situations (by us humans?), and quite another to say that, "therefore reality must be constituted of points". It is not even accurate to say that "quantification over points is indispensable" (say in physics); the full-fledged reduction already achieved of points to pluralities or sequences of regions simply belies that.

Thus, it seems to us that some combination of responses (2) and (3) is adequate regarding the theory of measure, and it is not necessary to take back any of our axioms.[6]

7.3.2 Differential geometry

This is an elaborate subject with many moving parts. Its central object is a *manifold*. In the abstract, one starts with an arbitrary set, X; its differential structure is furnished by a system of charts, in effect coordinatizations via one-to-one mappings from powers $\mathbb{R}^n$ of the reals $\mathbb{R}$ into X (with the stipulation that compositions between one and the inverse of another be smooth (C^∞)). Crucially, $\mathbb{R}$ is conceived as a point–set in the standard way, throughout the whole development. Because of its role as the lingua franca of modern space–time physics, differential geometry is clearly of central importance in any comprehensive study of continuity.

There are two contrasting ways of viewing this subject in relation to what has been accomplished here thus far: on the one hand, it is naturally regarded as key to the study of metrical structure of spaces of *variable* curvature, hence as generalizing the cases of constant curvature already treated (in Chapter 6). As such, one might propose an axiom system on, say, four-dimensional regions (of key importance in physics) along the lines of the systems described above, but somehow allowing for continuous variation of metrical structure, as in Einstein's general relativity.

Perhaps this can be done, and perhaps the effort would yield new insights. But we will not attempt that here, in light of the second view of the situation. As in the case of measure theory, just reviewed, no new axioms, beyond those of our one-dimensional regions-based theory, are really necessary. For, just as in the case of measure theory, the entire edifice of differential geometry as it stands really consists entirely of definitions, lemmas, and theorems. No axioms beyond those governing $\mathbb{R}$ itself, as a complete, separable, ordered field, are needed (see, for example, [90]).

It is also worth mentioning that the vast bulk of differential geometry can be carried out in one of the subsystems of second-order arithmetic of the Friedman–Simpson Reverse Mathematics program (see, for example, [114]), coupled with the fact that we have already assumed the equivalent of full second-order arithmetic, in the form of our key axioms of unrestricted comprehension and unrestricted fusions. We have

[6] On a related matter, Arntzenius and John Hawthorne [7] have shown how to account for continuous variation in a gunky background. The results are easily ported to the present theories. See also [59], §5.

already provided a thoroughgoing reduction of the Dedekind–Cantor theory of ℝ to the theory of the gunky line. Thus, it is in principle not necessary to come up with *sui generis* regions-based axioms for differential geometry.

7.4 Metaphysical Matters

7.4.1 *The questions; the perspective*

There are a number of different metaphysical issues in the vicinity, and we will only be concerned with some of those. First, there is the matter of whether space or space–time is physically real. The *substantivalist* follows Newton and says "yes"; the *relationalist* opposition follows Leibniz in saying "no". For the relationalist, all that is physically real are certain relations among physical objects, such as how far apart they are from each other (see Friedman [45] for an overview of the issue in terms of special and general relativity). The bits of the contemporary metaphysics literature that concern us here seem to presuppose substantivalism, at least for the most part.[7] Indeed, the theorists pose their questions in terms of "receptacles", those regions of space or space–time that can be exactly occupied by physical objects. If the relationalist is correct, then there are no such receptacles, since there is no independent space–time for them to be part of. So, as the metaphysical questions are usually posed, they make no sense to a relationalist. We will go along with substantivalism here as well, at least for the sake of argument. We do not know how to formulate the various questions concerning boundaries and contact in a relationalist setting (nor do we know how to reformulate our own, regions-based accounts, in relationalist terms).[8]

With substantivalism assumed, our main questions concern the *nature* of space or space–time. We will not address issues concerning whether it is Euclidean. We take that to be a matter for the relevant sciences. Our questions concern points. Most of the metaphysical literature presupposes (or seems to presuppose) that there is a fact of the matter whether space or space–time is punctiform—as noted, much of the literature just *assumes* that it is punctiform (and, indeed, assumes that space or space–time is punctiform as a matter of metaphysical necessity). We are not sure that there is a substantial question here, one of fact. Our perhaps tentative conclusion is that there is not.

Another preliminary concerns just what sort of "objects" we are talking about. Most metaphysicians are aware of the fact that physical objects in the actual world

[7] See, however, Josh Parsons [92]. Jonathan Schaffer [102] defends a view, called "supersubstantivalism", that *identifies* physical objects with regions of space–time.

[8] A relationalist account probably cannot restrict itself to relations among *actual* physical objects. What matters are the relations among *possible* physical objects. Intuitively, one might think that "possible objects" are gunky, in the sense that for every object, there could be one that is smaller. So perhaps a regions-based account of space or space–time might be modified to suit relationalism. Thanks to Paul Teller here.

(if there are any such objects) do not have the sort of boundaries envisioned in their metaphysical theorizing. They know, perhaps on authority from their science educations, that tables, baseballs, planets, and the like, are mostly empty or nearly empty space. Ordinary, garden-variety objects contain molecules that are themselves composed of atoms; and each atom has a nucleus around which electrons orbit (at least on the crude model). With quantum mechanics, perhaps understood second-hand, they know that sub-atomic particles like electrons simply do not have sharp boundaries. They are more like probability densities that fade off over space (at least on some understandings).[9] Contemporary metaphysicians also know that actual physical objects do not come into contact with each other in anything like the manner of their theorizing. There are forces that repel the objects long before they can actually touch each other (assuming that touching even makes sense for probability densities).

So what are these contemporary metaphysicians talking about, concerning boundaries, contact, and the like? There are two different orientations. Some of the literature explicitly takes itself to be exploring logical space (so to speak). They ask what boundaries and contact are like (or would be like) in those possible worlds where physical objects do have precise locations, and where objects do come into direct and full contact with each other (and not just "near contact", or overlapping probabilities, or something like that). As noted, members of this group are fully aware that the actual world is not one of those worlds; they profess interest in the other worlds.

One underlying theme seems to be that assumptions concerning mereology—the part–whole relationship—entail constraints on what location and contact might be. So, perhaps the best way to put this orientation is that the metaphysicians explore what is possible and what is necessary, given that such and such a mereological theory holds, and given that such and such intuitive principles are correct. They take themselves to be articulating entailments that illuminate connections between various mereological assumptions and the structure of space or space–time.[10]

A second orientation is that our questions are addressed in what P. F. Strawson ([117], 9) calls "descriptive metaphysics", a project of capturing "the actual structure of our thought about the world". Descriptive metaphysics aims "to lay bare the most general features of our conceptual structure", and not to revise this conceptual structure—and presumably not to figure out how the world really is.

Who is the "our" here? *Whose* thought is the target of this enterprise? Whose conceptual structure is the descriptive metaphysician after? As above, it is not that of the scientist, or the scientifically informed layperson; certainly not Penelope Maddy's [83] Second Philosopher. *That* theorist uses the findings of every relevant science to help her figure out what the world is like, including the structure of space–time, what

[9] Perhaps for these reasons, some argue that there are no tables, human bodies, and the like, a "mereological nihilism". Talk of tables is to be understood as talk of "simples arranged tablewise". See, for example, [124].

[10] Thanks to Gabriel Uzquiano here.

the nature of physical objects is, the extent to which such objects have determinate boundaries, and the ways they manage to interact with each other. Our metaphysicians do not take themselves to be doing *that*. Presumably, the metaphysicians are exploring "our" ordinary, everyday concepts of boundaries and contact. They want to figure out what is behind ordinary talk of whether a given stone has a rough or smooth surface, or an umpire's decision concerning whether a given baseball came into contact with a given bat, and thus whether it is a wild pitch or a foul ball.

One might take the relevant part of descriptive metaphysics to just be a study of the semantics of the English words "boundary", "surface", "touch", and "contact". In that case, we might get different results if we focused instead on different languages. Strawson [117], 9–10 writes that

> . . . up to a point, the reliance upon a close examination of the actual use of words is the best, and indeed the only sure, way in philosophy. But the discriminations we can make, and the connections we can establish in this way, are not general enough and not far-reaching enough to meet the full metaphysical demand for understanding.

The relation between Strawson's descriptive metaphysics and what Emmon Bach [9] calls "natural language metaphysics" is interesting, but would take us far afield to explore.

The two orientations—exploring logical space and articulating descriptive metaphysics—are sometimes combined. John Hawthorne ([52], 623) presents a thought experiment involving an infinite series of walls placed parallel to each other in a finite amount of space (and so all but a few of the walls are very thin, much thinner than the supposed diameter of any molecule). He anticipates an objection:

> Some may complain that the world is too distant to be worth being interested in. Actual walls do have extra repulsive forces, don't get to be rigid and impenetrable at any thickness, and so on. Such a reaction is far too hasty. Distant worlds can often be either revealing or therapeutic with regard to our actual conceptual scheme.

Hawthorne thus seems to hold that by examining what seems to happen in, or, better, what he and others are inclined to say concerning, bizarre, physically impossible situations—the playground of our first orientation above—we learn something about "our actual conceptual scheme", the main goal of descriptive metaphysics.

The enterprise thus seems to presuppose that there is such a thing as "our actual conceptual scheme", and that it is monolithic, consistent, and applicable in any and all conceivable situations. This presupposition is of a piece with what Mark Wilson [133] derides as the "classical picture", the view that the concepts we deploy are precisely delimited in all conceivable situations—that in any such situation, there is one and only one correct way to apply our concepts.

An opposing view begins with the observation that our "concepts" evolved to help us negotiate the world we find ourselves in, and they are applicable, at most, to the sorts of situations that come up here regularly. It would be something of a miracle if *any* of the everyday concepts we deploy could somehow apply smoothly, directly, and univocally to any conceivable situation whatsoever. As new situations arise, and as new deep theories of the world emerge, sometimes our concepts evolve. We sometimes have to improvise. That may take us into what Strawson calls "revisionary metaphysics" or, perhaps, Maddy's Second Philosophy.

Of course, we do not intend to resolve these matters of philosophical methodology here. With this background in place, let us turn to the issues at hand.

7.4.2 *Boundaries and contact*

Although we wish to avoid the question of whether to understand physical objects as three-dimensional or four-dimensional (or something else), it is tedious to keep repeating phrases like "space or space–time". For the sake of simplicity, we will phrase our questions and accounts in terms of three (or fewer) dimensions. Not much turns on this (or, to be precise, we will avoid matters that do turn on this).

As noted above, it is common to define a *receptacle* to be a chunk of space that a physical object can (exactly) occupy. If a given receptacle is occupied by a given object, then no part of the object lies outside the receptacle, and every part of receptacle is occupied by a part of the object. Of course, articulating this more precisely depends on what counts as a "part" of a physical object and, more central here, what counts as a "part" of a region of space. It seems to be supposed that each object has a unique receptacle.

Our first and, indeed, primary question concerns just which chunks of space are receptacles. On the present regions-based accounts of space, a natural thesis would be that any region is a receptacle. If we wish to avoid what Richard Cartwright [31] calls "scattered objects", like a chess configuration, a university campus, or the United States, then we can restrict attention to connected regions. If we wish to avoid objects that are infinitely large, relative to, say, a given stone, then we can restrict attention to bounded regions.

On the more standard Dedekind–Cantor accounts of space, regions are sets of points. So, from that common perspective, the task at hand is to determine which sets of points are receptacles. Is any non-empty set of points a receptacle? Or any Lebesgue measurable set (to avoid Tarski–Banach worries)? Or perhaps only open sets, or only regular open sets, or only closed sets, or only regular closed sets . . . are receptacles. Opinions vary.

Once we decide or figure out what the receptacles are, we then have to say something about what the boundaries of the objects are (if they have boundaries), and what it is for objects to be in contact with each other. Presumably, these questions are to be answered in terms of the receptacles the objects occupy. Consider, for example, an ice cube that exactly occupies a closed cube. Then, it seems, the object somehow includes its boundary. The ice cube itself has an interior and also a boundary. If, instead, an ice cube exactly occupies an open cube, then it—the ice cube—does not include its boundary. Every point occupied by the ice cube is contained in a neighborhood, say a sphere, that lies properly *inside* the ice cube, not on its edge.

Can we really make physical sense of, say, two ice cubes, one that contains its boundary and one that does not? Well, perhaps there are different kinds of objects, some of which contain their boundaries, and some of which do not. Questions like these become pressing if one thinks of space as composed of points in the now standard Dedekind–Cantor manner.[11]

When it comes to receptacles in a Dedekind–Cantor space, Hud Hudson [65] is the most liberal. He argues that *any* non-empty set of points can be a receptacle. Thus he envisions point-sized objects, scattered objects whose receptacles are finite or countably infinite sets of points, objects in 3-space that are planar squares, and the like. So, a fortiori, Hudson envisions open objects, those whose receptacles are open sets of points, and he envisions closed objects, those whose receptacles are closed sets of points.

Among other things, [65: Ch. 3] is devoted to an analysis of the notion of *contact*, of what it is for two objects to touch each other (and, it seems, for what it is for an object to touch itself). One preliminary proposal is that two objects are in contact just in case (a part of) the receptacle for one of the objects shares a boundary point with (a part of) the receptacle of the other. So two open spheres that are next to each other, separated by a single point, would qualify as in being contact, since the point between them is a common boundary—even though that point is not a member of either receptacle.

Against this, Hudson takes it as a criterion of "perfect contact" that "whenever two objects stand in that relation, it is not even possible that any two objects be more deserving of the description 'in perfect contact' than they". Our two open spheres fail that test. The spheres are separated by a point. Indeed, one can get an entire plane between them (and, on Hudson's view, one can get an entire *object* between them). Contrast this with an open sphere that is next to a closed sphere. These two objects are supposedly "more deserving of the description 'in perfect contact'" than the open spheres (not in terms of distance, of course), since we can't get *anything* to come between our second pair of spheres. So the early thought about contact is rejected.

Hudson then gives an extended series of thought experiments, involving things like "objects" that consist of a countable infinity of point-sized parts that converge to a point along a straight line (but does not include the limit point), along with zero, one, two, and three-dimensional objects that occupy the missing point. Hudson restricts attention to those objects which do not, and, indeed, cannot interpenetrate, in the sense that they do not, or cannot, share any parts, including point-sized parts. His considered definition, the fifth candidate, is this (p. 65):

[11] This is not to say that there are no issues concerning boundaries and contact on a regions-based account of space. We briefly address such issues below.

Necessarily, x touches y if and only if $\exists r_1, \exists r_2, \exists w, \exists v, \exists p$ (i) w is a part of x, whereas v is a part of y; (ii) w exactly occupies r_1, whereas v exactly occupies r_2; (iii) p is a boundary point of both r_1 and r_2; (iv) p is a member of exactly one of r_1 and r_2; and (v) $w \neq v$.

We would think that our everyday concepts of "contact" and "touching"—the relevant bits of "our conceptual scheme"—are just not equipped to deal with things like "physical objects" consisting of a countable infinity of point-sized parts. The everyday notions evolved to deal with things like rocks, baseballs, and bats, not such bizarre entities. Of course, one can always speculate as to how our everyday concepts could be, or should be, extended to cover such bizarre things, but one can also wonder whether there are unique ways to extend the notions. Hudson seems to presuppose that there is such a unique way to apply our notions to such "objects", or at least he takes that as a goal of the enterprise. One can also question the purpose of this piece of metaphysics. Why does it matter what we say concerning whether Hudson's "objects" are in contact or not? Our notions seem to work fine for the objects we encounter, in the normal range of circumstances in which we encounter them. Perhaps Hudson accepts Hawthorne's [52], 623 claim that "[d]istant worlds can often be either revealing or therapeutic with regard to our actual conceptual scheme". Presumably, Hudson takes the "distant worlds" here—those with point-sized objects—to be revealing.

Be that as it may, Hudson (ch. 3, §6) articulates a "knot" for which he has no satisfying resolution, by his own lights. Notice, first, that on Hudson's view, certain kinds of objects simply cannot touch each other (in the articulated sense of "perfect contact"). Two open cubical objects, for example, cannot share a boundary point occupied by one of them, for the simple reason that neither of them contains any of its boundary points. However, we can know how close two such objects can get to each other—they can get in position where only boundary points separate them. In contrast, consider two closed cubical objects. They also cannot touch each other, but for a different reason. If they did touch, they would have to have a *common* boundary point, but, since the cubes are closed, this point would have to be occupied by both objects, and so they would interpenetrate (a little). By Hudson's lights, this is not possible.

So here is the "knot". Suppose that, say, two closed cubical objects are moving toward each other at a constant velocity. Suppose they are oriented so that a side of one of them is always parallel to a side of the other. Let the two objects start out exactly 10 meters from each other, each moving towards each other at 1 meter per second relative to the other. What happens as we approach the five-second limit? The objects cannot end up touching, since they cannot touch (as a matter of metaphysical necessity). But no matter how close they get, they can always get closer without bumping into each other.

So what does happen? Of course, in the actual world, no object can exactly occupy a closed region. Moreover, the particles that make up each object exhibit repulsive forces that prevent them from touching (in Hudson's metaphysical sense of "touching"). But here we are (supposedly) exploring metaphysical possibilities. Hudson explains:

> . . . with the exception of pairing a closed object . . . with an open object . . . , any other pairing can be so oriented that both interpenetration and touching are not options; the items paired must . . . either slow down, jump, stop, turn, or otherwise deviate from their courses . . . Given the discussion above, we can clearly explain *that* such deviant behavior will occur, but just what explains *why*? . . . *Why* must approaching objects change speed or direction? I don't know—that's why the title of this section advertises a knot rather than a knot untied. (p. 82)

Clearly, Hudson encounters this "knot" only because of his ultra-liberal views on receptacles and, thus, on what kinds of objects are possible. Other metaphysicians are not as liberal as Hudson in allowing just any set of points to constitute a receptacle—a region that can be exactly occupied by a physical object. But at least some of the difficulties associated with points and such extend beyond this liberal orientation. Indeed, many such views face versions of Hudson's "knot".

Achille Varzi [125] provides a nice, mathematically informed instance of descriptive metaphysics, focused on matters of contact and boundaries. The article opens thus:

> The world of everyday experience is mostly a world of physical things separated in various ways from their environment: things with surfaces, skins, crusts, boundaries of some sort. These may not always be sharply defined, or so some would argue. Clouds, for instance, seem to have hazy boundaries, . . . But, generally speaking, boundaries seem to belong to the palette of basic ontological tools with which we commonly describe the middle-size reality of ordinary experience. . . . This general picture is so natural and pervasive that it is hard to deny boundaries a central place in our conceptual scheme. Boundaries are intrinsically connected with whatever entities they bound—in a way boundaries are even more important, since they are the first and primary things with which we seem to be directly acquainted. At the same time, the folk theory of boundaries is far from complete, and indeed several puzzles threaten any attempt to provide a comprehensive and systematic account of this notion. For instance, common sense is unaccustomed to the point-set topological distinction between open and closed entities. But then, how are the spatiotemporal relations between an extended object and its complement to be explained? And how is the relation of contact between contiguous substances to be accounted for? (p. 26)

Much of Varzi's article deals with what he calls "Peirce's puzzle": "Which color is the line of demarcation between a black spot and a white background? Does the line belong to the spot, or to the background?" (p. 29). If one wants to take ordinary talk of boundaries seriously, and literally, questions like these (supposedly) demand answers. One can give the boundary to only the spot, or to only the background, or one can think of the boundary as somehow shared, as belonging to both of them, or else one can say that it belongs to neither of them. As Varzi puts it, take "any entity *x*. Does the boundary (or any piece thereof) belong to *x* or to [the complement of *x*]? Does [the boundary] inherit the properties of the object or those of the complement?"

Varzi argues that at least some instances of this question are straightforward:

... we may find good reasons to maintain that material bodies such as stones or soap bars are the owners of their boundaries—their surfaces, in fact. Thus, where the complement meets an object of this sort, it will be open. (The object, in turn, will be the closed complement of the complement.) We may also ... argue that immaterial bodies such as holes are not the owners of their boundaries: these belong to the material bodies that host the holes. Thus, where the two meet, the complement (host) is closed and the entity (hole) open.

So let's start with a body of water, with respect to the air that surrounds it, an instance of our problem that Varzi attributes to Leonardo da Vinci. Is the boundary part of the water or part of the air? Suppose the boundary goes with the water. Now consider a block of wood floating on the water, partially submerged. According to Varzi's intuition, the block contains *its* boundary. Since the block is closed, the parts of the water that touch the block are open (there), while, as above, the parts of the water that are in contact with the air are closed. So, it seems, throwing the block on top of the water causes some of the boundary points of the water to be moved away (to make room for the boundary points on the block of wood). Of course, a similar issue confronts the option of holding that the water is open and the air closed. There we would wonder about the parts of air that are in contact with the wood.

There are also analogues of Hudson's "knot". Imagine two objects, both of which include their boundaries, say a baseball and a bat. Suppose the ball is thrown toward the bat; the bat is swung and makes contact with the ball, which then goes off toward the outfield. What happens to the boundaries of the bat and the ball at the moment, or during the period, when they come into contact? It follows from a suggestion that Varzi makes that a bit of at least one of the boundaries is temporarily pushed aside so that the boundary of the other can fit in that place.

Varzi concedes that some of these issues are difficult, perhaps irresolvable, but he suggests that this is not due to the notion of contact:

I think we may respond that such dilemmas are real, and yet insist on keeping the theory as it is. For, we may argue, the actual ownership of a boundary is not an issue that the theory must be able to settle. The theory only needs to explain what it means for two things to be connected. It doesn't need to give a full explanation of the underlying metaphysical grounds ... By the same token, we can say that every instance of Peirce's puzzle ... is truly problematic and yet extrinsic to our present concerns. Give me a theory of black spots, and make sure to tell me who gets the boundary—the spot or the background. Give me a theory of events, and make sure to tell me which gets the boundary—the movement or the rest. If we accept this response, we have a way of disposing of the puzzle in its general form ...

We respectfully submit that it might be better to adopt a view of space, time, and thus of receptacles, on which problems like this do not arise. Of course, Varzi is correct that any complete descriptive metaphysics should have something to say about surfaces and boundaries, or at least about our everyday *thought* and *talk* of surfaces and boundaries, but perhaps this can be done without having to worry about which objects are open and which closed (see below). As we have been emphasizing, *that* distinction simply makes no sense on the regions-based accounts of space.

Part of the problem here may be the thesis, or presupposition, that the boundary of a closed object is made up of the same kind of stuff as the rest of the object. So, for example, both the boundary and the interior of a (closed) block of wood consists of uncountably many point-sized bits of wood (pretending that this makes sense, at least in some possible worlds). Dean Zimmerman [134], 20, provides a thought experiment, an instance of Hudson's "knot", designed to show that two closed objects cannot share their boundaries. It seems to be based on the foregoing thought about the constitution of boundaries:[12]

> Take two closed cubes, for example, which move continuously toward one another until they touch by having faces which occupy the same plane. What prevents their motion from continuing, so that the cubes pass right through one another? Since each cube is the mereological sum of a set of two-dimensional objects lying in planes parallel to the coincident surfaces, and since these internal two-dimensional objects differ from the coincident surfaces only in being inside the cubes, why can *they* not coincide with two-dimensional parts of other objects as well? If extended objects are made up out of simples, and distinct simples can be in the same place at the same time, then does it not follow that any two distinct extended objects could be in the same place at the same time?

Some metaphysicians, notably Francisco Suárez ([118], disp. 40), Franz Brentano [25], and Roderick Chisholm (e.g. [33]; [34], ch. 8), hold that every extended, three-dimensional object, has a boundary consisting of point-sized parts, and that two such objects are in contact with each other when (and only when) at least one boundary point of one *coincides* with a boundary of the other.[13] Suárez and Brentano, at least, also held that, as Zimmerman [135], 158, put it, "the three-dimensionally extended parts of a thing are not made up out of indivisibles alone but also contain some 'atomless gunk', a substance all of whose parts have proper parts." The *boundary points* of different objects can coincide—they can be in the same place at the same time—but the *interior, gunky parts* of (at least some) different objects cannot coincide. So, in the above thought experiment, from Zimmerman [134], 20, the two cubical objects would continue moving until "they touch by having faces that occupy the same plane". At that instant, some of the points on one of the touching faces coincide with points on a face of the other one. The motion would then stop (or perhaps the objects would bounce and move apart), since the interior, gunky parts of the two objects cannot coincide.

[12] A simple is something that has no proper parts (i.e. no parts that are not identical to itself). In Dedekind–Cantor space, the only non-empty simples are singleton points. A gunky space has no simples.

[13] Zimmerman [134], especially §III, is an illuminating account of the relevant views of Suárez and Brentano. Barry Smith [115] contains a formal rendering of Chisholm's views on boundaries.

So far, so good, we suppose. That resolves these instances of Hudson's "knot". However, there are further issues concerning what we may call "interior boundaries". Consider any cubical object, say an ice cube. There is a plane surface within the object that separates its left half from its right half (in a given orientation). According to Suárez, Brentano, and Chisholm, that plane is an *interior boundary*, and it also consists of points (or point-sized parts). Indeed, there are point-sized parts of the object located *everywhere* inside it—along every plane that can divide the object. Presumably, there are continuum-many such interior potential boundary points. And for Suárez and Brentano, and, possibly also Chisholm, all of that is in addition to the gunky parts of the cube.

Suárez held that there is only *one* point-sized part of the cube at each interior spatial point. Zimmerman [134], [135] argues that this thesis has awkward, and, indeed, preposterous consequences. Suppose our cubical object is cut along a plane, to produce two separate half-cubes. The separated parts each have a face that was interior to the original object. Where did the point-sized parts of those two faces come from? One possibility is that the points on the original, internal plane constitute the surface of one of the pieces, and a new boundary—a new batch of point-sized parts—is *created* for the other. That seems weird, and arbitrary. It leaves the question of determining which half gets the point-sized pieces from the original, erstwhile internal plane and which gets the new boundary. Suárez, apparently, thought it best to maintain that the original internal plane is *destroyed* by the break, and *two* new surfaces are created (see Zimmerman [135], 159).

Brentano and Chisholm held, instead, that each internal spatial point of each extended object is occupied by *infinitely many* (perhaps even continuum many) co-located point-sized parts, one for each way the object can be divided into parts through that point. So suppose the object is cut along a given plane, and let p be a point on this plane. In the original object, there are infinitely many (indeed, continuum-many) point-sized parts of the object all located at p. When the object is cut, some of those point-sized objects go to one of the pieces, to be part of that new surface, and the rest of the point-sized objects at p go to the other piece.[14] Suppose that one of the pieces is cut again along another plane. Then the point-sized parts (along the original plane) are split again, some going to one of the new pieces and some to the other.

Brentano complained that the mathematician and philosopher Bernard Bolzano "was led . . . to his monstrous doctrine that there would exist bodies with and without surfaces, the one class containing just so many as the other, because contact would be possible only between a body with a surface and another without." Bolzano's

[14] One might soften the blow a bit by following Aristotle and insisting that the internal boundaries are only "potential", as in Chapter 1. See, for example, [115]. Varzi [125] does not countenance internal boundaries at all, even potentially. He suggests instead that when an object is cut, its boundary somehow stretches along the cut, to form the boundary surfaces of the separated parts.

"monstrous doctrine" is, of course, what gives rise to Hudson's "knot". Apparently, monstrosity is in the eye of the beholder.

As conceded, advocates of the present, regions-based account of space should also have something to say about boundaries and contact, or at least about ordinary talk of boundaries and contact. In a few places above, we show how to define "points" as certain equivalence classes of "convergent" sequences (or sets, or pluralities) of regions (in a few different ways). Then "lines" can be defined as equivalence classes of pairs of distinct points. It is just as straightforward to define "planes" in a three-dimensional regions-based space, as equivalence classes of non-colinear triples of "points". And so on for higher dimensions. So one option for an advocate of a regions-based space is to define the surface of a region as a certain set of "points". For example, on one definition, a convergent sequence *s* of regions is on the boundary of a region *r* just in case every member of *s* overlaps *r*, and no member of *s* is part of *r*. This resembles (or, better, is an instance of) Alfred North Whitehead's [128], 7, method of "extensive abstraction".

This, of course, makes boundaries into highly abstract objects. Officially, a boundary is a set of sets of sequences of regions. Varzi [125] argues that this does not make sense of ordinary talk of surfaces and such. For example, one might say that one *sees* the surface of a block of wood or that the surface of a given car has been *painted red*. Surely, we do not *see* sets of sets of sequences, and we cannot paint such things either ([82] notwithstanding).[15]

So, for some purposes having to do with descriptive metaphysics, it may be better to avoid reliance on this Whiteheadian "extensive abstraction". For what it is worth, we see little harm in speaking, informally, of the edges, surfaces, and boundaries of regions. In the treatment above, we engaged in discourse like this quite a bit. For example, in presenting the two-dimensional theories, we speak of the top, bottom, left, or right side of a given generalized quadrilateral (even before we defined points and lines), although we would often put those locutions in scare quotes. Of course, we insist that edges and the like are not *parts* of regions, in any sense of that term. Our Aristotelian mantra is that the only parts of regions are other regions. We'd go further and insist that for the purposes of descriptive metaphysics, there is no reason to reify edges and boundaries at all; no reason to think of them as *things*, or at least as things that are independent of the objects they bound. To adapt an Aristotelian thesis, the right side of a generalized quadrilateral can no more be separated from the generalized quadrilateral than the smile of the Cheshire cat can be separated from the cat. In speaking, informally, of the sides or edges or boundaries of a region, we are just calling attention to certain aspects of the region, for example, where it is located with respect to other regions.

[15] We presume that contemporary advocates of the foregoing point-based metaphysical views agree that one does not literally *see* zero or one-dimensional things either. Can light reflect off a point or a line? Can paint stick to a single point? How? Does a paint-point somehow get stuck on a wood-point? How?

From a regions-based perspective, contact between objects is particularly easy to understand. Two objects touch if the regions they occupy are contiguous, in Aristotle's sense. Again, regions are contiguous just in case there is a direction in which no region lies between them. Consider, for example, an object *A* that exactly occupies a generalized quadrilateral, and another object *B* that exactly occupies the right quadra-extension of that generalized quadrilateral. Clearly, *A* is in contact with *B*: the right edge of *A* is snug against the left edge of *B*. If we define "places" and "edges" via "extensive abstraction", then we can say that these two "edges" are located in the same "place": the convergent sequences along the right edge of *A* are, in fact, exactly the same as the convergent sequences along the left edge of *B*. But there should be no temptation to worry about co-located things (at least not here). The right edge of *A* is the right edge *of A*, and is not to be thought of as an object somehow floating free of the "rest" of *A*, and the left edge of *B* is similarly tied to *B*.[16] If we don't reify "edges", there is at least less temptation to think otherwise.

In an earlier work, Hudson [64], 84–9, launches an argument against the very (metaphysical) possibility of "material atomless gunk". That is, he argues that there *must be* "material simples", physical objects that have no proper parts. Hudson allows that there are (or might be) two different kinds of material simples, "pointy objects", those that exactly occupy a single point of space or space–time, and "maximally continuous objects".[17] Present concern, of course, is with the former. Hudson notes that the following five theses are (obviously) inconsistent with each other:

(1) The "Doctrine of Arbitrary Undetached Parts": if a material object x exactly occupies a region r, and if r^* is any "exactly occupiable subregion" of r, then x has a material object part that exactly occupies r^*;
(2) "Necessarily, no hunk of material atomless gunk exactly occupies a point-sized region";
(3) "Necessarily, any hunk of material atomless gunk exactly occupies some region or other";
(4) "Necessarily, any region has at least one point-sized subregion"; and
(5) "Necessarily, any point-sized region is exactly occupiable".

Hudson supports (5) on the basis of a premise that material simples are point-sized objects, together with "the claim that material simples are possible". It seems to us that this just begs the question. Someone who doubts that there are, or that there can be,

[16] Thanks to Sarah Broadie for suggesting something along the present lines in the context of Aristotle's conception of "contact" and "contiguity". It might be noted, however, that the resolution suggested here may not be entirely Aristotelian. Aristotle does wonder what happens at certain temporal boundaries. Consider, for example, a moment when someone begins walking. Is there a first instant (or "now") of motion or is there a last instant of rest?

[17] Following Markosian [84], Hudson defines an object x to be "maximally continuous" if "x is a continuous object (neither spatially nor temporally gappy), and there is no region of space–time, s, such that (i) the region occupied by x is a proper subset of s, and (ii) every point in s falls within some object or other". This seems to presuppose that space–time is itself punctiform.

point-sized objects will also doubt that "material simples are possible" (if all material simples are point-sized).[18] Hudson suggests that "the most promising strategy for the gunk theorist is to take aim at premise (4) on the grounds that 'gunky space' is possible" (p. 89). Hudson notes that

> [r]ecent and intriguing defenses of a Whiteheadian theory of space—a view that might deserve the description of "gunky space"—are available in the literature . . . [S]trictly speaking, questions about the occupiability of point-sized regions could not even arise, for there would be no point-sized regions to have questions about.

The present study is, of course, a "Whiteheadian" account of space (or space–time), or at least it can be expanded to become one. Apparently, Hudson thinks that we violate a metaphysical law. He concludes:

> The mathematical project of constructing points out of sets of infinitely many, converging, nested, extended regions does not guarantee the metaphysical possibility of gunky space any more than the formal consistency of geometries of arbitrary many dimensions establishes the metaphysical possibility of four-dimensional space . . . More pressing, however, it seems to me that the claim that "space is gunky" must have its truth value as a matter of necessity. But then, if we ground our belief in the possibility of material atomless gunk with an appeal to gunky space, we will effectively rule out the possibility of material simples . . . And that consequence, I submit, is too high a price to pay. (pp. 89–90)

One might retort that the fact that one can describe space as punctiform, along standard Dedekind–Cantor lines, does not guarantee that space is punctiform, as a matter of metaphysical necessity (or possibility). Who has the burden of proof here? We are not metaphysicians, and we don't put forward a thesis that space or space–time is really gunky. Presumably, the actual structure of space or space–time is a scientific matter or, perhaps better, it is up to scientists to decide which description of space or space–time best suits their needs. We take ourselves to have demonstrated that the various Euclidean and non-Euclidean spaces can be described, mathematically, either way: the regions-based and punctiform theories are mutually interpretable (and, indeed, definitionally equivalent).

Hudson informs us that the "cost" of holding that space is really gunky is "too high", but we are not given much of an economic analysis. Above, when we mentioned Brentano's comment on Bolzano, we suggested that what counts as "monstrous" is in the eye of the beholder. Apparently, the same goes for what is "too high a price to pay". For what it is worth, our own "intuition" is that eschewing the very possibility of point-sized objects is not really a cost at all, at least not on the metaphysical front, let alone a high cost—but perhaps that just begs the question.

[18] As Hudson notes, denying premise (2) would saddle a metaphysician with point-sized objects that have point-sized proper parts, a dire conclusion (supposedly).

7.4.3 Grounding and fundamentality (and junk)

Arntzenius [5], 225–6, argues, tentatively, that space or space–time is gunky. He claims that this is a far-reaching idea (see also Arntzenius [6], 126):

It is standardly assumed that space and time consist of extensionless points. It is also a fairly standard assumption that all matter in the universe has point-sized parts. We are not often explicitly reminded of these very basic assumptions. But they are there . . .

The alternative is that space and time and matter are 'pointless', or 'gunky'. The idea here is not that space and time and matter have smallest finite-sized bits, that space and time and matter are 'chunky'. Rather the idea is that every part of space and time and matter has a non-zero, finite, size, and yet every such part can always be subdivided into further, smaller, parts. That is to say, the idea is that every part of space and time and matter has a non-zero size, and yet there is no smallest size.

Let me emphasize how radical this idea is. It is very natural to think that any thing decomposes into some ultimate collection of fundamental parts. And it is very natural to think that the features of any object are determined by the way that object is constructed from its ultimate parts, and by the elementary features of these ultimate parts. Indeed, much of the history of science can be seen as an attempt to break down complex objects and processes into ultimate parts, and to find the laws that govern these ultimate parts. But if there are no smallest regions, and if there are no smallest parts of objects, then a spatial or temporal decomposition of a region, and of an object, cannot bottom out at an ultimate level. The idea that the features of large regions and large objects are determined by the features of minimal-sized regions and minimal-sized objects cannot work if space and time, and the objects in it, are gunky, i.e. pointless. Space, time, and objects would simply not have ultimate parts. There would just be an infinite descending chain of ever-smaller parts. A somewhat dizzying prospect.

As noted above, more than once, the thesis that space and time are gunky is not new. Aristotle held this view, and it is implicit in Euclid's *Elements*. Throughout the ancient, medieval, and early modern periods, there were perhaps a few voices defending the view that space, time, and matter are punctiform, but they were a distinct minority—and were thoroughly criticized by the mainstream mathematicians and philosophers. The main alternatives were varieties of atomism, views that describe space or matter as "chunky", as Arntzenius puts it, and the Aristotelian, gunky picture, following the dominant mathematics and science of the various periods.

Of course, the thesis that space and time, or space–time, is punctiform is now the dominant orthodoxy, to the extent that the view is sometimes called "classical" (perhaps due to its reliance on classical logic). This can be attributed, of course, to the overwhelming success of the Dedekind–Cantor accounts of continuity and the real numbers. And *that* is due, in large part, to the power of actual infinity—to the "paradise" that Cantor bequeathed us.

But there is no evidence, so far as we know, that Aristotle, Euclid, and the medieval and early modern mathematicians and scientists, found gunky space and gunky time to be a "dizzying prospect", nor was this a "radical" idea. Indeed, for most of our history, it was the received view. Here, for example, is Leibniz [72], 162 (quoted in [20] and [21]):

For, although there are atoms of substance, namely monads, which lack parts, there are no atoms of bulk, that is, atoms of the least possible extension, nor are there any ultimate elements, since a continuum cannot be composed out of points. In just the same way, there is nothing greatest in bulk nor infinite in extension, even if there is always something bigger than anything else, though there is a being greatest in the intensity of its perfection, that is, a being infinite in power.

Presumably, the supposed "very natural" thought that "any thing decomposes into some ultimate parts" was also not held very much throughout this history, since it is inconsistent with the dominant Aristotelian/Euclidean account of space, and the natural account of matter that goes with that. To be sure, Aristotle and many others did hold that some things are to be explained in terms of others, and that the latter are to be explained in terms of yet other things, and so on, until one hits a fundamental level. Bolzano's "ground-consequence" relation and Frege's hierarchy of "proof" and "justification" are later articulations of this idea (see, for example, [108]). The key thesis, in all three cases, is that the grounding (or explanatory) relations are objective, not a matter of how this or that person comes to know or comes to discover anything. Contemporary metaphysicians also often speak of what is "fundamental", again taking that notion to be objective. For the historical figures, however, and for at least some of the contemporary metaphysicians, it is a hierarchy of *propositions*, and not of objects or substances or anything like that. For example, it may be that the *axioms* of Euclidean geometry are fundamental; not to be understood in terms of anything else. That much is quite consistent with space, time, and objects being gunky. Indeed, the (fundamental) axioms might declare it to be so.

Against this, at least some contemporary metaphysicians, such as Jonathan Schaffer [102] and Theodore Sider [113], do hold that grounding and fundamentality apply, first and foremost, to objects. That is, some *objects* are more fundamental than others, and some are absolutely fundamental, not grounded in anything else.

From that perspective, someone may indeed find the possibility of gunk a "somewhat dizzying prospect". However, as Schaffer points out, this assumes that the more fundamental objects are the *smaller* ones: a is more fundamental than b only if a is a *part* of b. Arntzenius seems to hold this as well, with the "very natural" thought that "any thing decomposes into some ultimate collection of fundamental parts" (assuming that the "decomposing" is mereological). Against this, Schaffer points out that the "more fundamental" relation, among objects, might be somewhat independent of the mereological relations among those same objects. That, at least, would make the world safe for (the metaphysical possibility of) gunk.

There is a closely related matter. Recall that, in contemporary, metaphysical jargon, "junk" is a sort of opposite to gunk: a space is junky if each object is a proper part of another object. The main "semi-Aristotelian" theories developed in this study are not junky. Indeed, recall that those theories all have an unrestricted principle of composition: given any regions, there is a (unique) fusion of just those regions. One version of this is a scheme:

$$\exists u\Phi(u) \rightarrow [\exists x\forall y\{y \circ x \leftrightarrow \exists z(\Phi(z) \,\&\, z \circ y)\}],$$

where Φ is any predicate of the language lacking free x.

It follows that there is a single region that is the fusion of all regions—just instantiate $\Phi(x)$ with the formula $x = x$. Of course, that region is not a proper part of any other.

In Chapter 3, we develop a more Aristotelian account of space, at least in one dimension. There, the principle of unrestricted composition is replaced with a thesis that any pair of regions has a fusion (and, thus, any finite collection of regions has a fusion). That space is junky, and well as gunky—such spaces are sometimes called "hunky" (see [20]).

Schaffer [102], §2.4 argues against the metaphysical possibility of hunk and, ultimately, of junk. This would be interesting news to, say, Aristotle and Euclid, and most of the major figures throughout the medieval and early modern periods (see the above passage from Leibniz). Schaffer's argument turns on the relation of grounding and the concomitant property of being fundamental. If we hold that the grounding relation is well-founded—as a matter of metaphysical necessity—then it must "bottom out" (or "top out") at objects that are fundamental, objects that are not grounded in any others. It would be ad hoc to think that middle-sized objects are the most fundamental. So either objects are grounded in their parts or they are grounded in objects of which they are parts. Schaffer then argues that one of those disjuncts must hold of metaphysical necessity. The former option—that objects are grounded in their parts—is inconsistent with gunk (as per the above argument from [5]). But Schaffer takes it that gunk *is* metaphysically possible. So all objects are grounded in objects of which they are parts. So there can be only one fundamental thing, the object which is the fusion of everything. It is the One or the Universe. Thus, there can be no junk.

Of course, there are ways to resist this conclusion. One is to reject the notion of objective grounding relations altogether. Why think that the universe comes ordered in such a pleasing manner, as if it was designed to make it easy for us to understand? Or, as above, one can hold that the grounding relation holds between *propositions*, not between objects. It would follow, if grounding is well-founded, that certain *propositions* are fundamental, not grounded on any others. But this, by itself, has no ramifications for the mereological structure of space, time, space–time, or objects.

Another argument against junk begins with a thesis that the above unrestricted principle of composition is itself a necessary truth. Again, it would follow that there is a region that is the fusion of all regions, and that region cannot be a proper part of anything (given the other axioms of mereology). Apparently, some contemporary metaphysicians do hold that unrestricted composition is a metaphysical necessity (for example, Lewis [77], Armstrong [4], and Sider [112]).[19] So those metaphysicians are committed to the metaphysical impossibility of junk (and thus of hunk)—bad news for Aristotle, Euclid, and Leibniz.

If we start with an infinite set (or sequence) of regions and take their fusion, then we are, in effect, treating the set as an actual infinity. Indeed, since we produce a *single* region, we have to think of the given set, or sequence, as somehow total, or complete. As above, a common theme throughout most of our history was that there simply is no actual infinity. The only coherent notion, it was claimed, is that of the potentially infinite. As discussed in Chapter 1, Cantor argued for the exact opposite of this, that the potentially infinite only makes sense in terms of the actually infinite:

> I cannot ascribe any being to the indefinite, the variable, the improper infinite in whatever form they appear, because they are nothing but either relational concepts or merely subjective representations or intuitions (imaginationes), but never adequate ideas. ([27], 205, note 3)

> . . . every potential infinite, if it is to be applicable in a rigorous mathematical way, presupposes an actual infinite. ([28], 410–11)

This attitude is now dominant in mathematics and science (with the exception of constructivists). If one takes the Cantorian perspective to be a metaphysical necessity, then perhaps unrestricted composition is also a metaphysical necessity, and so junk is impossible, contra Aristotle, Euclid, Leibniz, and just about every other major mathematician and philosopher before, say, 1820. We leave it to the gentle reader whether to do a modus ponens or a modus tollens on this conditional.[20]

7.5 Verbal Disputes

There are less "liberal" views on receptacles that avoid the above Hudson–Varsi knots concerning boundaries and contact. We show here that in several such cases—the most plausible in our opinion—the resulting view of space, or space–time, is essentially a notational variant of one of our (semi-Aristotelian) regions-based accounts.

[19] See Bohn [20], [21], [22] for discussion.

[20] It might be added that contemporary metaphysicians are not consistent on what they take unrestricted composition to be. Some dub the weaker principle that any two (and thus any finite number of) things have a fusion as "unrestricted" composition. Trenton Merricks [86], 8, for example, writes that "unrestricted (universal) composition" is "the thesis that any two things compose something", although he cites "defenders" of this who actually defend the stronger, unrestricted principle, or at least a thesis that every *set* of objects has a sum (e.g. Cartwright [31]; Lewis [77], 212–13; Leonard and Goodman [75]). Sider [113], 176, says that "If composition had been unrestricted (that is, if it had been the case that for any x and y there exists a mereological sum of x and y)" (see also p. 119). However, in this regard, Sider cites Dorr [38], 234, who writes that *Universalists* hold that for "any things whatsoever, no matter how scattered and miscellaneous they might be, there is something they compose". The latter has no restriction to finite fusions or even set-sized fusions. Sider [112], 7, cites the full version: "the principle of unrestricted mereological composition" is that "for any objects, there exists such a thing as the mereological sum, or fusion of those objects". Of course, it is only the fully unrestricted principle of composition that is inconsistent with junk, and that notion is bound up with actual infinity.

Recall that, in topology, a set is regular open if and only if it is identical to the interior of its closure. Such sets are sometimes called "open domains". A set is regular closed if and only if it is identical to the closure of its interior; such sets are sometimes called "closed domains".

Cartwright [31] argues that a set of space (or space–time) points is (or can be) a receptacle if and only if it is non-empty and regular open. On that view, there are no substantial puzzles concerning contact: two objects are in contact just in case their receptacles share a boundary point. Equivalently, objects are in contact if there is no open set—no receptacle—that lies between them. Talk of boundaries is straightforward, except that it is insisted that no receptacle, and thus no object, includes its boundary. Thus the Varsi-type worries concerning "Peirce's puzzle" (from section 7.4.2) do not arise.

Gabriel Uzquiano [121] articulates two "moderate" views on receptacles, both of which, he argues, avoid some serious difficulties that apply to more "liberal" views on receptacles (such as those of Hudson [65]). One of the moderate views is the Cartwright thesis that only non-empty regular open sets are receptacles, and the other is that only non-empty regular closed sets (closed domains) are receptacles. Contact is also straightforward on the latter view: two objects are in contact just in case their receptacles *share* a boundary point. Concerning Peirce's puzzle, on Uzquiano's second moderate view, each object includes its boundary. Of course, that view has it that distinct objects in contact do inter-penetrate—they share a part of their boundaries. However, the view is consistent with a thesis that the *interiors* of (some) objects cannot inter-penetrate. Perhaps this is enough to satisfy intuitions against inter-penetration.

Zimmerman [134] provides two options for receptacles, in a Dedekind–Cantor space. Both are consistent with his argument that physical objects are constituted, at least in part, by gunky material. One option is that all receptacles are open (which, as above, makes the notion of contact relatively simple). But, surely, from Zimmerman's perspective, it is not the case that *every* non-empty open region is a receptacle (unless an object can have more than one receptacle)—Hudson's intuitions notwithstanding. Consider, for example, the set of points interior to a given sphere, except for its center and fifteen other interior points. That set is open, but surely Zimmerman would not consider it to be a receptacle (or at least not the only receptacle for a given object). Zimmerman's second option is that all receptacles are closed, together with an account of boundaries and contact along the lines of Brentano and Chisholm (i.e. where boundaries consist of multiply co-located point-sized parts; see section 7.4.2). But, again, surely Zimmerman would agree that not every non-empty (extended) closed set counts as a receptacle. Consider a closed sphere together with sixteen other scattered points, or a closed sphere together with a line segment exterior to it. It seems to us that, from Zimmerman's perspective, only *regular* open regions, or else only *regular* closed regions, should count as receptacles. From his perspective, receptacles should not have isolated boundaries, nor isolated point-sized holes internal to the object, the very things ruled out by regularity.

Now recall that there are two topological models for the foregoing regions-based theories, in a standard Dedekind–Cantor punctiform space. In one of these models, "regions" are regular open sets of points. The fusion of a set of "regions" is the interior of the closure of the union. In the other topological model, "regions" are regular closed sets of points, and the fusion of a set of "regions" is the closure of the interior of its union.[21] These are in fact the notions of fusion adopted by Uzquiano.

Of course the authors discussed here are not concerned solely with pure mathematics, but with theories of *space* (or space–time) and physical objects, or at least the spatial (or spatio-temporal) regions that physical objects can occupy (in some possible worlds, or else "on our conceptual scheme"). Still, the tight connection between the punctiform and the regions-based theories at least suggests that neither Cartwright nor Uzquiano nor, arguably, Zimmerman needed to begin the deliberation with a Dedekind–Cantor account of space, at least not for their accounts of receptacles, boundaries, and contact. They would have essentially the same account(s) of receptacles, boundaries, and contact if they had begun with a regions-based theory of space, since, for them, the receptacles are or might as well be the regions in the present gunky spaces.

Zimmerman [134] seems to recognize something like this, and even seems to favor a regions-based account of space. Early in the article, he writes

> But if space were not made out of points, spatially extended objects could not be made out of point-sized simples. Where would the simples be if there were no point-sized locations for them to occupy? By adopting the analytical, punctual approach to the geometry of physical space, I merely afford the view I intend to refute a running start. (p. 2)

Later, when discussing Whiteheadian accounts of receptacles, Zimmerman writes, "Whitehead's treatment of space obviously does yield the result that, if there were extended objects, then they would have no simple parts—there being, on Whitehead's view, no simple regions in which to put them" (p. 18, note 26). This seems to presuppose that every object has a receptacle in space or space–time.

Arntzenius [5] (see also [6], ch. 4, and Russell [101]) characterizes a gunky space. To formulate the account, Arntzenius asks the reader to start with a standard, Dedekind–Cantor space and put on "blurry glasses" that cannot see the boundaries of regions. More formally, define two sets of points to be equivalent just in case they differ only on boundary points. In particular, if a point p is in one of them but not the other, then p is a boundary point of the latter. It is straightforward to show that this is indeed an equivalence relation. In effect, if two sets are equivalent, then we cannot tell them apart when wearing the blurry glasses. In Arntzenius's model, the gunky regions are isomorphic to the equivalence classes.

Arntzenius then notes that each equivalence class contains exactly one regular open set of points, and exactly one regular closed set of points. So, like Cartwright and

[21] Borgo [23] notes that regular sets form a natural model for a number of point-free geometries.

Uzquiano and, arguably, Zimmerman, Arntzenius would have the same account of space if he had just began with the regular open sets (or the regular closed sets). And, as above, it would be the same account of space if he had begun with one of the foregoing regions-based theories.[22]

We can get a similar result via a different route. Suppose one holds that space is punctiform but that objects are gunky. This is natural for those who accept a Dedekind–Cantor account of space (or space–time), but also hold that any object can be divided (thus rejecting point-sized objects, contra Hudson). Then, perhaps, one might also hold that each object "exactly occupies" more than one set of points. On such a view, for example, it seems natural to say that a gunky ice cube exactly occupies both an open and a closed cube, as well as any other set that contains the interior of the cube and is contained in its closure. Let S be the set of sets of points that can be "exactly occupied" by an object. Define the following binary relation on S: a set a is related to a set b just in case any object that exactly occupies a also exactly occupies b. Arguably, this is an equivalence; indeed a is related to b just in case they differ only on boundary points, i.e. just in case the closure of a is identical to the closure of b. Then, as with the Arntzenius construction, each equivalence class contains exactly one regular open set and exactly one regular closed set, and the equivalence classes are isomorphic to the foregoing regions-based space. So our theorist might as well have begun the deliberation with a regions-based space.[23]

Perhaps there is no substantial dispute between advocates of a gunky space and advocates of a Dedekind–Cantor punctiform space. Perhaps the dispute is verbal. Recent years have seen a proliferation of literature on the nature of verbal disputes. We conclude with an application of two such accounts to our case.

The main agenda-item of Eli Hirsch's "Physical-object ontology, verbal disputes, and common sense" [63] is an argument that many of the ontological disputes in contemporary metaphysics are merely verbal. One such dispute is over whether it is the case that, for any objects whatsoever, there is a mereological sum of those objects. For example, is it the case that there is an object that is the mereological sum of LeBron James's favorite jersey and a given chicken in Ohio? Some metaphysicians say "yes"; others say "no". Hirsch's thesis is that this dispute is merely verbal.

To support this, Hirsch proposes a test for verbal disputes. An example, to motivate the test, is someone, A, who insists that standard drinking glasses are cups. The rest of us do not agree with A, of course. Intuitively, the "dispute" here *is* merely verbal; it comes down to the meaning of the word "cup". Hirsch asks us to "imagine a linguistic community which is in all other ways as close as possible to our actual

[22] As noted in Section 7.3.1, Arntzenius is not satisfied with this gunky account of space (see also [101]). One reason is that it does not support a countably additive measure. As noted, he develops a second account of space, motivated by a different pair of "blurry glasses", one that cannot distinguish sets if the difference between them has Lebesgue measure zero. The result is a gunky space with a straightforward, countably additive measure, but with some pretty weird regions.

[23] Thanks to Eli Hirsch for suggesting this account of space and objects.

English-speaking community but in which everyone agrees with *A*." Call this the *A*-community and call the language they speak *A*-English. Hirsch argues that the most plausible (or charitable) way to interpret *A*-English is to say that, in that language, the extension of the word "cup" includes standard drinking glasses.

Let us assume, plausibly, that as the word is understood in ordinary English, the extension of "cup" does not include standard drinking glasses (at least in typical contexts). Speakers of ordinary English and speakers of *A*-English would have no trouble understanding each other, once they realize that they speak (slightly) different languages. It would be routine to translate the languages to each other. The *A*-English word "cup" would be rendered "cup or drinking glass" in ordinary English, and the ordinary English word "cup" would be rendered "cup that is not a drinking glass" in *A*-English. We find no factual disagreements between members of the two linguistic communities. As Hirsch puts it, the dispute between *A* and the rest of us is merely verbal because one can fully understand what is going on in this way.

Hirsch concludes:[24]

> The general characterization of a verbal dispute is one in which the controversial sentences are most plausibly interpreted as having different truth conditions in the different languages associated with the contending positions, so that each position is correct in its own language . . . The verbalness of a dispute may turn on the reference of a term such as "cup", or it may turn on the meaning of logical constants—as in a verbal dispute as to whether "Either John or Mary is ill" is true when both John and Mary are ill—or any semantic or syntactic feature of language that affects the truth conditions of sentences. ([63], 72)

To apply Hirsch's test to the present case, imagine two scientists. One of them, *DC*, adopts a Dedekind–Cantor punctiform space, or space–time, and the other, *Gu*, uses a (semi-Aristotelian) regions-based account, along present lines. We can have them both holding that space and space–time are Euclidean, or else they can both adopt some other metric (so long as they agree on that). As emphasized, each will have no trouble "translating" the other's language into her own (after proving a few easy theorems). When *Gu* speaks of a "region", *DC* will interpret that as a statement about

[24] Hirsch has so-called "coarse-grained" truth-conditions in mind: the corresponding sentences should be true in the same possible worlds. There seems to be a family resemblance between this test for verbal disputes and the aforementined notions of mutual interpretability and definitional equivalence, from section 7.2. The Hirsch test applies to (interpreted) *languages*, while mutual interpretability and definitional equivalence apply to (formal) *theories*. To get fanciful, suppose that two languages L_1, L_2 pass the Hirsch test, and consider a speaker *s* who speaks both languages and knows that the Hirsch test is met by them. Let T_1 be the set of sentences in L_1 known by *s* (including her semantic knowledge, setting aside the specter of paradox), and let T_2 be the set of sentences in L_2 known by *s* (again including semantic knowledge). Then, it seems, T_1 and T_2 are mutually interpretable. Moreover, let Φ be a sentence in one of the languages; let Φ' be the result of translating Φ to the other language; and let Φ'' be the result of translating Φ' back to the first language. Presumably our subject *s* knows that Φ and Φ'' have the same (coarse-grained) truth-conditions, and so knows that they are equivalent (in the respective language/theory). So T_1 is definitionally equivalent to T_2.

a regular open set of points (or else a regular closed set of points). When *DC* speaks of a "point", *Gu* renders that as an equivalence class of convergent sets (say).

As above, the connection between the respective "languages", via the translations, is extremely tight. The set of points, in any model of the Dedekind–Cantor space, is isomorphic to the corresponding set of equivalence classes of sets in any model of the regions-based theorist's theory. And the set of regions for the regions-based theorist is isomorphic to the set of regular open (and also to the set of regular-closed) sets of Dedekind–Cantor points. Indeed, each of them can recognize the isomorphism (either going via the translation or else "going native" and construing the other's theory in ordinary model-theoretic terms). As above, the two *mathematical* theories are definitionally equivalent. So the translations preserve truth-values for any statement purely about space (or space–time). Our question is whether this extends to the entire language. It clearly extends to straightforward claims of physics or ordinary talk of regions or objects.

Consider the modal pronouncements of our disputants. Suppose that both of them hold that the question of whether actual space (or space–time) is gunky is contingent. That is, they agree that there are gunky worlds and that there are punctiform worlds. *DC* adds that the actual world is punctiform, while *Gu* disagrees, insisting that the actual world is gunky (although it is hard to see what sort of evidence either might marshal for this). Then, it seems, the dispute fails the Hirsch test. Or, at least, it is hard to see how the test might be met. We get similar results if one of them holds that space (or space–time) has its structure of necessity while the other holds that it is contingent.

So let us just assume that *DC* and *Gu* each assert that their account of space (or space–time) holds of metaphysical necessity.[25] This seems to be a common view among metaphysicians: the geometry of space or space–time might be a contingent matter, to be decided in physical theory, but not its mereology. Recall, for example, Hudson's [65], 89–90, assertion, "it seems to me that the claim that 'space is gunky' must have its truth value as a matter of necessity". On this combination of views, the foregoing translations extend to any statements concerning the mereology of space (or space–time). For example, *Gu* says that, necessarily, every region has a proper part. *DC* renders that as the statement that, necessarily, every regular open set of spatial (or space–time) points has a proper regular open subset. Conversely, *DC* says that every occupied region has a point as a part. *Gu* renders that as a statement that every set of equivalence classes of sets of nested regions of a certain sort has an equivalence class as a member (so that they do not mean the same by "part").[26]

[25] Thanks to Eli Hirsch for pointing this out. It is, we think, a bit curious that we get this dispute to be verbal only if the disputants make metaphysically strong claims.

[26] We suppose that *DC* might assert that it is possible for space itself to consist of a single point (or a finite set of points). It is hard to see how *Gu* might render that possibility. Let us assume instead that *DC* holds that space or space–time has the full Dedekind–Cantor structure, as a matter of metaphysical necessity.

We turn next to our theorists' metaphysical pronouncements, as per the previous section. *DC* might claim that talk of points (or points themselves) is (or are) *fundamental*, or perhaps that it is *more fundamental* than talk of regions. *Gu* would assert the negation of this, insisting that talk of regions (or regions themselves) is the more fundamental. Or *DC* might say that only points are "real", with regions "derived" or "constructed" from them; *Gu* would say the opposite. The obvious move here would be to suggest that the theorists do not mean the same by phrases like "fundamental", "real", "derived", and "constructed". Each might render the other's use of those terms with respect to the type-hierarchy (with respect to the translated sentences). So *Gu* interprets *DC*'s claim that points are more fundamental than regions as a theorem that equivalence classes of sets are members of sets of such equivalence classes.

To be sure, one or the other of our theorists might insist that they are using "fundamental" in a metaphysically robust sense, aimed at the underlying structure of the universe, the one true way to carve nature at its joints (following Sider [113]). If this meta-metaphysical ploy can be sustained, and if these last pronouncements cannot be plausibly rendered in the other's language, then the dispute is not verbal (and neither is any other dispute over what is fundamental). We have nothing further to contribute here, leaving the reader to decide whether to do a modus ponens or a modus tollens.

Let us turn to what our theorists might say concerning what sorts of physical objects are possible, and how physical objects relate to each other. *DC* might say that it is possible for there to be point-sized objects (say, following Hudson [65], as above); such objects are exactly located in regions consisting of a single point. Her counterpart *Gu* can say what it would be for an object to be point-sized: the object is contained in every member of a convergent set of regions (and it has no receptacle). We suppose that *Gu* could countenance such objects, at least as a metaphysical possibility. But what of *DC*'s claim that every object is exactly located in a region? As we have seen, Zimmerman [134] suggests that proponents of regions-based spaces have trouble here: "Whitehead's treatment of space obviously does yield the result that, if there were extended objects, then they would have no simple parts—there being, on Whitehead's view, no simple regions in which to put them" (p. 18, note 26). We see no insurmountable problem here. *Gu* could render *DC*'s thesis as "necessarily, for every object *o*, there is a unique set *s* of convergent sets of regions, such that *o* overlaps every member of every member of *s*".

Instead, and, we think, more plausibly, *DC* may adopt a less "liberal" view of receptacles, as above, following Cartwright, Zimmerman, or Uzquiano (and perhaps Arntzenius and Russell). Then point-sized objects are ruled out, perhaps as a matter of metaphysical necessity. As noted, in such cases, the theories of receptacles are close indeed—they are isomorphic. Moreover, it is even clearer that the Hirsch test for verbal disputes is met. The overall theories seem to be notational variants of each other. The theorists also seem to be on a par with respect to defining fields, measures, or whatever. Such is the way with isomorphic structures.

Like Hirsch, David Chalmers [32] also attempts to explicate the notion of a verbal dispute, for the purpose of helping to resolve more general issues in philosophy. Chalmers suggests that "a broadly verbal dispute is one that can be resolved by attending to language and resolving metalinguistic differences over meaning" (p. 526). This sounds correct, but, of course, it is not always easy or straightforward to deal with language, nor is it always easy to figure out whether that is what we are doing.

Chalmers proposes a "heuristic for detecting and dealing with verbal disputes", which he calls "the method of elimination". The idea is that "one eliminates the key term, and one attempts to determine whether any substantive dispute remains". In more detail:

> To apply this method to a dispute over a sentence *S* that is potentially verbal with respect to a term *T*, one proceeds as follows. First: one bars the use of term *T*. Second: one tries to find a sentence *S′* in the newly restricted vocabulary such that the parties disagree nonverbally over *S′* and such that the disagreement over *S′* is part of the dispute over *S*. Third: if there is such an *S′*, the dispute over *S* is not wholly verbal, or at least there is a substantive dispute in the vicinity. If there is no such *S′*, then the dispute over *S* is wholly verbal (except in the special case of vocabulary exhaustion . . .). (pp. 526–7)

Chalmers illustrates his technique with the question of whether Lee Harvey Oswald acted alone, with respect to the word "alone". Once one bars the use of that term, it is likely, Chalmers suggests, that one can indeed find various sentences *S′*, which do not contain the word "alone", and such that a nonverbal dispute over *S′* is part of the original dispute.

Chalmers's method of elimination is a non-starter in the present case. We might start by banning the words "point" and "region", but it also seems that the participants may not mean the same by words and phrases like "part of", "overlap", and even "real" and "fundamental". So we would have to ban all of those. It seems that we do not have enough vocabulary left for our characters *DC* and *Gu* to differ. In other words, we seem to have here a rather straightforward instance of what Chalmers calls "vocabulary exhaustion": "If a language has a limited vocabulary, then it might be that once one bars a key expression, one can no longer even formulate any issue that might potentially resolve the original issue." (p. 530).

Chalmers argues that cases of vocabulary exhaustion are "rare", writing that in "a language such as English, these exceptions will arise only for terms expressing especially primitive concepts". One further problem here is that our disputants do not agree on what the "primitive concepts" are. Presumably, *DC* takes "point" to be a primitive (or close to one), while *Gu* does not, taking "point" to be defined in terms of regions. Conversely, *Gu* takes "region" to be a primitive, while for *DC* a region is a set of points.

Chalmers goes on to propose a "special case" of the method of elimination, which he calls the subscript gambit:

> Suppose that two parties are arguing over the answer to "What is *X*?" One says, "*X* is such-and-such", while the other says, "*X* is so-and-so". To apply the subscript gambit, we bar the term *X*

and introduce two new terms X_1 and X_2 that are stipulated to be equivalent to the two right-hand-sides. We can then ask: do the parties have nonverbal disagreements involving X_1 and X_2, of a sort that resolving these disagreements will at least partly resolve the original dispute? If yes, then the original dispute is nonverbal, and the residual disagreement may serve as the focus of a clarified dispute. If no, then this suggests that the original dispute was verbal (unless we have reached the point of vocabulary exhaustion . . .). (p. 532).

So what we would do here would be to develop a common language for our disputants. We would replace words like "point" and "region" (and "fundamental", "primitive", etc.) with subscripted counterparts. So we would have terms like "point$_{Gu}$" and "region$_{DC}$", "primitive$_{Gu}$", etc. That might help prevent confusion and misunderstanding, but it would not help much beyond that, since there does not seem to be much left for the parties to disagree over, given that they agree on the structural connections between, say "point$_{Gu}$" and "region$_{Gu}$". For example, both would recognize an isomorphism between points$_{DC}$ and equivalence-classes$_{Gu}$ of sets of regions$_{Gu}$. So the disagreement passes Chalmers's test as well as Hirsch's (under similar assumptions).

Chalmers, however, notes a limit to the methodology:

At a certain point, when the method of elimination is applied, a proponent may say "That's bedrock." Here, the thought is that we have reached a point where only cognate expressions can be used to state the issue and where there is no hope of finding a relevant disagreement at an "underlying level". Further application of the method will just lead to wheel spinning and ultimately to vocabulary exhaustion. (pp. 544–5)

Bedrock disputes are, it seems, not merely verbal. We imagine that, in the present case, the disputants would not agree as to what counts as "bedrock". It would be good to have more detail concerning when one, in fact, has a case of a bedrock dispute, other than the failure of the Chalmers tests.

Even if the present dispute passes the tests for being verbal, or *merely* verbal, it does not follow that there is no difference between our theorists, and nothing to help choose which one to follow. Sometimes language matters. The question is what it matters for.

Gu can lay claim to what Uzquiano [123] calls "mereological harmony": the mereological structure of space exactly matches the mereological structure of objects. *DC* cannot claim that, without countenancing point-sized objects. Of course, *DC* can render the statement of mereological harmony in terms of a complex relation that holds between the punctiform space and the gunky objects (on the less liberal views of receptacles). It might turn on which theory is, overall, more simple and elegant.

Along similar lines, perhaps the regions-based theorist will be less tempted to postulate point-sized objects (and then worry about contact between such), and she may be less tempted to reify boundaries, as being somehow physically real, and she may be less tempted to worry if there are both open and closed objects. Her account of space or space–time does not seem to suggest such possibilities (even if they are somehow logical possibilities). Of course, as we have seen the Dedekind–Cantor theorist can avoid these worries as well, say by settling on one of the less liberal views of receptacles.

From the other perspective, we noted above (section 7.3) that it at least seems simpler to capture things like Lebesgue measure, differential geometry, variable curvature, and perhaps continuous variation by working with points, or "points". One might conclude that if we need points anyway, then perhaps we should just start with them. For these purposes, it seems, the punctiform Dedekind–Cantor theories are simpler and more elegant. Arntzenius [6] argues that this gives a reason to think that points are real, in a metaphysically robust sense.

Along similar lines, David Bostock [24] provides a nice reconstruction of a gunky space, based on the ideas of Whitehead and Bertrand Russell. He concludes, however, that it might be better to start with points, on grounds of simplicity, or at least familiarity:

> I conclude that, while it is quite natural to take points and regions as equally basic, we should accept that there is an ontological economy in 'constructing' the one from the other. But the 'construction' can be done either way round. It is of interest that it can be done in the direction that interested Whitehead and Russell . . . But in these days our mathematicians and our physicists are much more familiar with the other direction, and there is surely no good reason for them to drop this approach in favour of the other.

To be sure, on pragmatic grounds alone, a scientist or anyone else will prefer a simpler and more elegant theory, if all else is equal. In the present case, arguably, all else *is* equal. That, indeed, is one of our main conclusions. Simplicity and elegance seem to be in the eye of the beholder, and also relative to one's underlying purposes. There is, of course, a vast and vexed literature on the nature of simplicity and elegance, and on its epistemic status. The central question is the extent to which simplicity and elegance bear on scientific or metaphysical truth.[27] We will not enter into that matter here.

[27] See the excellent survey in [10].

8

Scorecard

The purpose of this final chapter is to briefly assess a number of different accounts of the continuous, concerning the extent to which each sanctions intuitive features that have been attributed to continuous entities or substances throughout the ages. We consider eight different conceptions, or theories, and assess each on nine intuitive features. As with many matters, of course, short summaries can be misleading. Some cases are fairly straightforward, but for others, a bit of discussion is required, as it is not clear how the intuitive feature in question is to be articulated in that context.

The upshot is that no one account of the continuous gets it all; each forgoes at least some intuitive feature of the continuous, a feature is that has been thought to be essential to the concept. Our overall conclusion is that the notion of continuity is more like a cluster concept, one that is up for sharpening in various incompatible ways.

To quote the Rolling Stones:

> You can't always get what you want. But if you try, sometimes, well, you just might find, you get what you need.

In the case at hand, what you need depends on one's background and, theoretical goals. Those vary across time.

8.1 The Accounts

1. **Ancient atomism.** The atomists were Aristotle's major opponents on the relevant issues (see Miller [1982]). Michael J. White [1992] speculates that there are two versions of atomism. One is a view that material *objects* come in atomic pieces that cannot be divided or otherwise broken up; all of what we see is the result of atoms moving around in an otherwise empty space, colliding with each other. The more radical version is a thesis about *space* and *time* themselves, that there are atomic, extended regions or temporal intervals that cannot themselves be divided. We will be concerned with the radical version here, whether or not any ancient figure ever held such a view.[1] In effect, the radical atomist denies that space is continuous, and so it is not really an account of continuity. We include it here for the purposes of contrast.

[1] Arguably, Zeno's paradox of the stadium is aimed at this radical atomism. That gives at least some reason to think that the view was held. Returning to the present, some interpretations of quantum mechanics posit that there can be nothing smaller than a Planck length (about 1.6×10^{-35} meters) and

2. **Aristotle.** The view is covered, at least superficially, in Chapter 1. The mathematics is essentially that of Euclid's *Elements*, although we present a more modalized version in Chapter 3.

3. **Dedekind–Cantor.** This, of course, is now the received view. In the one-dimensional case, a point is, or corresponds to, either an equivalence class of Cauchy-sequences of rational numbers (Cantor) or a Dedekind cut in the rational numbers (Dedekind). Regions are nothing but sets of points. This perspective is sometimes called the Triumvirate view, the third member being Weierstrass.

4. **Non-standard analysis.** It invokes infinitesimals, and the background logic is classical.

5. **Intuitionistic analysis.** On at least some versions, a real number is an equivalence class of Cauchy sequences, as on the Dedekind–Cantor view, but such sequences are treated as only potentially infinite. The background logic, of course, is intuitionistic. See, for example, Beeson [1985] or Troelstra and van Dalen [1988].

6. **Smooth infinitesimal analysis.** A rather different account of the continuous, allowing nilpotent infinitesimals and requiring intuitionistic logic. See Bell [1998] or Kock [2006]. For discussion of the justification of intuitionistic logic in smooth infinitesimal analysis, see [58].

7. **Point-based predicative analysis.** See Feferman and Hellman [1995], [2000]. Also Hellman [2004].

8. **Point-free geometry and analysis.** This is the account developed here. There are three different versions: One is "semi-Aristotelian", and makes full use of classical logic and actual infinity (Chapter 2 above). A second is more Aristotelian, getting by with only potential infinity, and the third is predicative, lying between the other two (Chapter 3). We also develop a constructive version, from the superstructure (Chapter 4), but the results here are the same as those of intuitionistic analysis.

8.2 The Intuitions

A. **Infinite divisibility.** Is it the case that each extended region has a proper part, or can have a proper part? In one dimension, for example, is it the case that each interval can be bisected (or trisected etc.)? This seems to be the one feature that all of the accounts share, except, of course, radical atomism. On most of the views, however, infinite divisibility is not sufficient for continuity. To deny infinite divisibility is to hold that there are extended indivisibles.

B. **Archimedean property.** There are two formulations. (i) Let r and s be any two magnitudes, of the same dimension (e.g. two line segments). Then there is a natural number m, such that the result of adding r to itself m-times is larger than s. As

nothing shorter than a Planck time (about 10^{-43} seconds); or at least that those are the smallest meaningful units. That much, at least, is consonant with radical atomism.

Aristotle puts it (*Physics*, 8, 10): "By continually adding to a finite quantity, I shall exceed any definite quantity".

In Euclid, this Archimedean principle appears as a definition: "Magnitudes are said to have a ratio to one another which are capable, when multiplied, of exceeding one another" (Book 5, Definition 4).

Here is the second: Let r and s be any two magnitudes of the same dimension. Suppose we remove at least half of r, and then remove at least half of the remainder, etc. Continuing in this way, we will eventually produce a magnitude smaller than s.

Euclid (10, 1):

> Two unequal magnitudes being set out, if from the greater there be subtracted a magnitude greater than its half, and from that which is left a magnitude greater than its half, and if this process is repeated continually, there will be left some magnitude less than the lesser magnitude set out.

C. **Do points exist?** This will depend, of course, on just what a point is. Euclid says "a point is that which has no part". If "part" here is understood in terms of contemporary mereology, then the Euclidean definition should be that a point has no *proper* part; points are indivisible. A point should also have zero dimension, and it should lack positive size (or have zero size). In other words, a point is not extended. One might also think that a point is a *fixed location*, so that if p and q are points, it should be determinate whether $p = q$, and if $p \neq q$, it should be determinate how they are located relative to each other.

D. **Are points parts of regions? Are extended regions composed of (nothing but) points?**

E. **Viscosity; indecomposability** There is a long-standing belief that a continuous substance has a kind of unity. There is something that holds it together. As Aristotle puts it, in the *Physics*:

> The continuous is just what is contiguous, but I say that a thing is continuous when the extremities of each at which they are in contact become one and the same and are (as the name implies) contained in each other. Continuity is impossible if these extremities are two. This definition makes it plain that continuity belongs to things that naturally, in virtue of their mutual contact, form a unity. (227a6)

Let us call this (alleged) feature of continuous things *viscosity*. The notion seems to be aimed at physical *objects*; it is not clear what it means for a *region* to form a unity. One possible consequence of viscosity, which might apply to regions is *indecomposability*. The idea is that it is not possible to cleanly divide an object or region cleanly into parts. Any attempt to do so will either leave something out or create something new (such as a boundary). Formally, a one-dimensional space is indecomposable if for any interval A it is not the case that there are regions B, C, which are discrete—having no part in common—and whose sum or union is A.

F. **Intermediate value theorem.** This is easiest to state for accounts formulated in terms of real numbers, or things enough like real numbers. For any continuous function f if there is a number a such that $f(a) < 0$ and if there is a number b such that $f(b) > 0$, then there is a number c (between a and b) such that $f(c) = 0$. A related principle is that if two continuous curves (such as a circle and a line, or two circles) cross each other, then they must meet somewhere—typically there is a point that lies on both. This feature relates to one of the so-called gaps in Euclid's *Elements*.

G. **Cauchy completeness.** Does every Cauchy sequence converge?

H. **Dedekind completeness.** Does every bounded set have a boundary?

I. **Infinity and infinitesimals.** First, are there actually infinite regions, perhaps relative to a given unit region? Equivalently, are there extended regions that are infinitesimal relative to another? Second, is it the case that for every extended region, there are extended regions (of the same dimension) that are infinitesimal relative to it? This last, of course, is the denial of one version of the Archimedean property. These matters may not be essential to continuity, but our theories do differ on them.

8.3 Ancient Atomism (Radical Version)

This is the one account in our list that denies infinite divisibility (A). The atomic bits of space and the atomic intervals of time are extended and cannot be divided further. That is what makes them atoms (or atomic regions or intervals). Infinite divisibility, in some form or other, is the one feature that is shared by all of the rival accounts of continuity. Of course, this version of atomism denies that space and time are continuous, in any intuitive sense.

For what it is worth, atomism is consistent with an Archimedean property (B), assuming it makes sense to multiply regions by natural numbers. The view is perhaps also consistent with some regions being infinite relative to others. Atomism denies that points exist (C), at least as points are usually understood. To be sure, atomic regions and temporal intervals are indivisible (by definition), and thus they meet the Euclidean definition of having no (proper) parts, but they are extended; each atom has positive size, and so does not qualify as a point.

The atomist denies that points are parts of regions, since she denies that points exist at all (D). The atoms, at least, display viscosity (E). They are unities, in every sense of the term. We also have indecomposability for the atoms; since they are indivisible, they cannot be divided cleanly into parts. Of course, any non-atomic regions or temporal intervals are decomposable into their atomic constituents.

The intermediate value theorem (F) does not apply, or, if it does, it is vacuous. According to the atomist, there simply are no continuous curves, and so there is not much point in wondering what happens when one continuous curve crosses another one.

Something similar holds of Cauchy completeness (G). A Cauchy sequence is one whose elements get arbitrarily close to each other. Presumably, the atomist simply denies that there are any such sequences, and thus there is no sense in figuring out if they converge. It is similarly difficult to see what to make of Dedekind completeness for this view (H).

The question of whether there can be infinite regions (I) would come down to whether there are, or can be, an actual infinity of spatial atoms. We won't speculate on this.

8.4 Aristotle

As we saw, Aristotle holds that spatial and temporal regions are infinitely divisible (A), and he endorses the Archimedean property (B). These are reflected in Euclid's *Elements* as well.

Aristotle does recognize the existence of points (C), as does Euclid. For Aristotle, points are, primarily, the *limits*, or endpoints, of line segments.[2] Each point is metaphysically dependent on the segment that it bounds, and cannot be considered apart from the segment that it bounds.

For Aristotle, the points interior to a segment exist only potentially, not actually. An interior point represents a place where a segment *could be* divided. But unless the segment *is* divided there, the point does not exist—it is not actual. On each line segment, there is a potential infinity of such points.

The central feature of Aristotle's views here is that regions (in any dimension) are not composed of points (D). In contemporary terms, for Aristotle, space and time are gunky:

> Now if the terms "continuous", "in contact", and "in succession" are understood as defined above—things being "continuous" if their extremities are one, "in contact" if their extremities are together", and "in succession" if there is nothing of their own kind intermediate between them—nothing that is continuous can be composed of indivisibles: e.g. a line cannot be composed of points, the line being continuous and the point indivisible.
>
> (*Physics*, Book 6, 231a1)
>
> . . . since indivisibles have no parts, they must be in contact with one another as whole with whole. And if they are in contact with one another as whole with whole, they will not be continuous: for that which is continuous has distinct parts . . . as we saw, no continuous thing is divisible into things without parts . . . Moreover, it is plain that everything continuous is divisible into divisibles that are infinitely divisible. (231b)

As we saw, Aristotle explicitly says that continuous things are "unities", so we have viscosity (E). As noted, however, it is hard to see what the mathematical fallout of this may be. What formal feature corresponds to this? One suggestion is what we call indecomposability, the idea that a continuous thing cannot be divided cleanly into parts. Aristotle endorses a somewhat attenuated kind of indecomposability. When we divide a line segment, we *create* something new, the endpoints of the resulting pieces. It might be more accurate to say, instead, that when a line segment is broken, then

[2] Also for Aristotle, line segments are just the sides of plane figures, and plane figures are just the faces of (actual or possible) physical objects.

two points that had previously only existed potentially are made actual. As he puts it (*Metaphysics* 3.5 (1002a28–b11)):

> For as soon as bodies come into contact or are divided, the boundaries simultaneously become one if they touch and two if they are divided. Hence, when the bodies have been put together, one boundary does not exist, but *has* ceased to exist, and when they have been divided, the boundaries exist where they did not exist before (for the point, being indivisible, was not divided into two).

As we will see, the theories based on intuitionistic logic (intuitionistic analysis and smooth infinitesimal analysis) exhibit a much stronger kind of indecomposability, one that is, in a way, the opposite of the Aristotelian one.

We know of no textual evidence for (or against) a version of the intermediate value theorem (F). As noted, Euclid seems to presuppose something like this, at least for circles and lines. This is one of the so-called "gaps", and there is no reason to think that Aristotle would balk at this as well.

It is probably too much of an anachronism to wonder whether Aristotle endorsed, or would have endorsed, either Cauchy completeness (G) or Dedekind completeness (H). Cauchy sequences and Dedekind cuts are both infinitary objects. Aristotle, along with just about every major mathematical thinker until the modern period, rejected the actual infinite. Instead, he endorsed the potentially infinite. Those are processes, like divisibility, that can be iterated indefinitely. There was no sense made of a *theory of* potentially infinite things, like Cauchy sequences and Dedekind cuts, something formalizable in a language with variables ranging over those. So it is hard to ask whether Aristotle would have held that every Cauchy sequence has a limit—presumably an actual limit—and whether every Dedekind cut has a bound—presumably an actual bound.

Finally, Aristotle held that all geometric objects are finite (I). There are no infinite lines, planes, etc. Euclid does not have a principle that rules these out, but infinitary objects are never used. Instead, he speaks of what happens if a given line is extended. The Archimedean property rules out regions that are infinitesimal relative to finite regions.

8.5 Dedekind–Cantor—The Triumvirate

This one is perhaps the most straightforward in our lineup. Extended regions are themselves uncountably infinite sets of points, and the parts of a given set are its (non-empty) subsets. So each extended region has a proper part (A), and the space is Archimedean (B).

Points clearly exist (C). A point is either an equivalence class of Cauchy sequences (Cantor) or corresponds to a Dedekind cut (Dedekind). Equivalence classes of Cauchy sequences, for example, do have "parts", namely their members, and each Cauchy sequence has "parts", namely its elements. So, strictly speaking, Euclid's definition of a "point" as that which has no parts, does not apply. But on either formulation, points do not have spatial or temporal or otherwise geometric parts. And each point is or represents a determinate location in the space.

On the Dedekind–Cantor view, points are indeed parts of regions (D). Indeed, since, as noted, regions just are sets of points, regions are composed of nothing but points.

It may be something of a category mistake to ask about viscosity (E). There is some sense in which a set is a "unity"; it is one thing after all. But there does not seem to be anything that holds the set together. Indecomposability fails for the simple reason that every subset has a complement—such is classical logic. Let m be any interval, say an open interval (r, s), and let t be any number such that $r < t < s$. Then m can be divided into the half-open interval $(r, t]$ and the open interval (t, s). Those two regions have nothing in common; and their union is m. Viscosity/indecomposability is probably the most central intuition that is lost on this account. Of course, an open interval cannot be decomposed into two open intervals, nor can a closed interval be decomposed into two closed intervals, but that is a far cry from full indecomposability.

The intermediate value theorem holds (F), and the space is both Cauchy complete (G) and Dedekind complete (H). There are infinite regions, in the sense of being infinitely large; the entire space is one such. But no extended, bounded region is infinitesimal relative to another extended, bounded region (I).

8.6 Non-standard Analysis

Some of the answers depend on the exact formulation of non-standard analysis. Let us begin with a language containing a name for each real number and each set of real numbers. The key feature, of course, is that any statement in the first-order language that is true of the Dedekind–Cantor real numbers is also true here. So some of the features carry over, namely those that correspond to first-order statements in the given language.

The space is infinitely divisible (A), for the same reason that the Dedekind–Cantor one is: extended regions are uncountably infinite sets of points.

The Archimedean property fails, however (B). Let r be any finite number and let t be any infinite number. Then, in the one-dimensional case, the intervals $(0, r)$ and $(0, t)$ are a counterexample to the Archimedean property. Non-standard analysis yields a prime example of a non-Archimedean field.

Points exist (C); points are parts of regions; and extended regions are composed of nothing but points (D). Indecomposability fails (E), in the same sense as in the Dedekind–Cantor account. The intermediate value theorem holds, at least for functions that can be defined in the extended language (F).

It is not clear what is to count as a Cauchy sequence here. If we take a sequence to be a function from the *standard* natural numbers into the space, and if we define such a sequence to be Cauchy if, for every *standard* rational number $r > 0$, the values of the sequence are eventually all within r of each other, then there are Cauchy sequences that do not converge. However, if we define a sequence to be any function from the "natural numbers" of the space to the space, and give the usual definition of Cauchy sequence in terms of the "rational numbers" of the space, then the space is Cauchy complete (G).

Dedekind completeness fails (H): the infinitesimals are bounded and they have no least upper bound. Of course, every first-order definable set that is bounded has a least upper bound.[3]

And, of course, there are infinitely large regions, and plenty of infinitesimals (I). Indeed, for every interval i containing more than one point, there are intervals (containing more than one point) that are infinitesimal relative to i.

8.7 Intuitionistic Analysis

As noted, a real number is an equivalence class of Cauchy sequences of rational numbers, but such sequences are regarded as only potentially infinite. We do have infinite divisibility (A) and the space is Archimedean (B), at least in the negative sense that no finite intervals are infinitesimal relative to each other.

The matter of points is a bit more subtle (C); it depends on what a "point" is. We can put Euclid's definition aside, for the same reason we did above. Equivalence classes of Cauchy sequences and Cauchy sequences themselves have parts. But a real number has no *geometrical* parts; nothing is "smaller" than the singleton of a real number. So intuitionistic real numbers are "points" in that sense.

However, one might think of a point as something like a fixed or determinate location. At least some real numbers qualify for that. Consider, for example, the sequence $\langle 1, 1, 1, \ldots \rangle$. That seems like a location on the number line. But consider the Cauchy sequence corresponding to the decimal expansion of the real number π: $\langle 3, 3.1, 3.14, 3.141, \ldots \rangle$. The sequence is regarded as potentially infinite; we cannot "pass to the limit". In a sense, we never reach the limit. The terms of the sequence *approach* a determinate location, but *is the sequence itself* a determinate location? We do not know how to answer this question. It almost seems out of place.

Above, we also suggested that, on some construals, if p and q are points, it should be determinate whether $p = q$, and if $p \neq q$, it should be determinate how they are located relative to each other. That fails here, rather badly. Identity is not decidable in intuitionistic analysis.

Our next question is whether points are parts of regions, and whether extended regions are composed of nothing but points (D). If we take a "point" to be a real

[3] This is because every first-order sentence that is true in the standard real numbers ($\mathbb{R}$) is also true in the non-standard numbers. So suppose that we have that $(\exists y)(\forall x)(\Phi(x) \rightarrow x < y)$ (where Φ is first-order). Then, in the standard real numbers, the Φ's have a least upper bound. So the Φ's have a least upper bound in the non-standard realm as well.

number, and a region to be a collection of real numbers (as in the Dedekind–Cantor scheme), then, of course, "points" are parts (subsets) of regions, and regions are composed of nothing but these "points". But the "if" is at least questionable.

Indecomposability holds, in a strong sense (E). No interval can be divided, cleanly, into two parts.[4] We saw that, for Aristotle, any attempt to break a continuous entity creates something *new*—the boundaries. Here, any attempt to divide an interval will *leave something out*. The metaphor is that if you cut an interval, something will stick to the knife. So connected regions display a kind of unity.

By way of compensation, however, the intermediate value theorem, as it is classically stated, fails (F). Indeed, that statement of the intermediate value theorem entails excluded middle.

Intuitionistic analysis is Cauchy complete: every Cauchy sequence of real numbers is equivalent to a real number (i.e. a Cauchy sequence of rational numbers). However, it is not Dedekind complete (H).

Finally, the intuitionist follows Aristotle in rejecting the actual infinite (I). And there are no non-zero infinitesimals.

8.8 Smooth Infinitesimal Analysis

Smooth infinitesimal analysis is presented axiomatically (in Bell [1998]). Recall that unlike Robinsonian non-standard analysis, the logic of smooth infinitesimal analysis must be intuitionistic, on pain of inconsistency. Different models of the theory have different features. We do have infinite divisibility (A) in all models, at least for finite regions that have positive size. On at least some natural definitions, the space has Archimedean (B) models, again in at least the negative sense that no two finite intervals of positive size are infinitesimal relative to each other (see Bell [1998, 108]). It also has non-Archimedean models.

The elements of smooth infinitesimal analysis are not constructed, as in intuitionistic analysis. Rather they are postulated, via the axioms. An element has no parts, geometric or otherwise, and so the elements meet Euclid's definition of "point" (C). As above, however, one might think of a point as something like a fixed location, so that excluded middle should hold for points. There are some such "points", such as 0 and 1. But, as with intuitionistic analysis, identity is not decidable. So it is not refutable that some of the elements are too "amorphous" to serve as points, in that sense (see Bell [2006, 318–21]).

Our next question is whether points are parts of regions, and whether extended regions are composed of nothing but points (D). If we take a "point" to be an element of the space, and a region to be a collection of elements, then, of course, "points" are

[4] The reason is that in intuitionistic analysis, all functions on an interval are (uniformly) continuous. If a given interval could be decomposed non-trivially into two parts, x, y, there would be a function on the interval that assigns 1 to the members of x and 0 to the members of y. Such a function would not be continuous.

parts (subsets) of regions, and regions are composed of nothing but these "points". But if we think of "points" as fixed locations—where excluded middle holds—then regions are not composed entirely of points (although any extended region will contain points).

Indecomposability holds, in a strong sense (E). No interval can be divided, cleanly, into two parts, since all functions are smooth and thus continuous (see note 4). Again, any attempt to do so will leave something out. Moreover, the intermediate value theorem fails (F) (see Bell [1998, 105–6]).

We do not know about Cauchy completeness (G). The space is not Dedekind complete (H). Indeed, the nilsquares are bounded, but they have no least upper bound.

One can talk about the entire space as a region, and so, in that sense, there are actually infinite regions (relative to an interval) (I). There are also models of smooth infinitesimal analysis which contain "invertible" infinitesimals, with multiplicative inverses (like those of non-standard analysis).

8.9 Predicative Analysis

As with the Dedekind–Cantor view, here extended regions are infinite sets of points, and the parts of a given set are its non-empty subsets. So each extended region has a proper part (A), and the space is Archimedean (B). Points exist, in the same sense that they do in the Dedekind–Cantor view (C), and regions are nothing but sets of points (D). The space is not viscous, and regions are fully decomposable (E). The background logic is classical, and so any given sub-region has a complement. The space is Cauchy complete (G), in the sense that every predicatively defined Cauchy sequence converges (G), and it is Dedekind complete in the sense that every predicatively definable set that is bounded above has a least upper bound (H). The same goes for the intermediate value theorem (F). There are infinitely long regions, but no infinitesimal ones (I).

8.10 Point-free Geometry and Analysis

All of our theories display infinite divisibility (A): there is an axiom that every region has a proper part, an interval in the one-dimensional cases, a rectangle in the two-dimensional ones, etc. The Archimedean property (B) is a theorem in the semi-Aristotelian and the predicative versions of the theory, and it is an axiom in the Aristotelian one. In the latter, the Archimedean property is stronger: for any interval i and every region m, there is a natural number n such that m is part of the interval obtained by extending i on its left and its right n times. This is possible since, for the Aristotelian, all regions are finite in extent.

The systems do not have any points (C), at least not in any primary sense. In all three of them, one can define "points" to be equivalence classes of some sets of regions, following Whiteheadian "extensive abstraction", although this goes beyond Aristotelian resources. But, clearly, "points", so defined, are not parts of regions (D). "Points" are certain sets. Clearly, it is not the case that extended regions are composed of nothing but points. They are not composed of anything except gunky regions.

There is perhaps a sense in which each region is a kind of unity, but we do not have indecomposability (E). In the one-dimensional case, it is a theorem of the semi-Aristotelian theory and an immediate consequence of an axiom of the other two that any interval can be bisected. So intervals can indeed be divided cleanly. Indeed, this is a virtue of the semi-Aristotelian theories. The same goes for many regions in spaces of higher-dimension.

The intermediate value theorem (F) does hold in the two-dimensional, Euclidean theory (which is also semi-Aristotelian), using the definitions of "line" as certain sets of "points". Recall that that theory is notationally equivalent to the Dedekind–Cantor one.

The semi-Aristotelian theory is both Cauchy complete (G) and Dedekind complete (H), once those notions are properly formulated. The Aristotelian theory is neither, or at least that cannot be proved. Indeed, that theory has countable models. The predicative one is Cauchy complete, but not Dedekind complete in the full, classical sense.

Finally, none of the theories has a region that is infinitesimal relative to an interval, and analogously for the higher-dimensional cases (I). That is a consequence of the Archimedean property. The semi-Aristotelian and the predicative version have infinite regions. The whole line is one such. The Aristotelian theory does not, insisting that all regions are finite in extent.

	infinite divisibility	Archimedean	points exist	points are parts	viscosity/ indecomposability	intermediate value theorem	Cauchy completeness	Dedekind completeness	infinity/ infinitesimals
ancient atomism	no	consistent	no	no	yes	not applicable	not applicable	not applicable	consistent
Aristotle	yes	yes	as endpoints	no	yes (weak form)	assumed	not applicable	not applicable	no
Dedekind–Cantor	yes	yes	yes	yes	no	yes	yes	yes	yes/no
non-standard analysis	yes	no	yes	yes	no	yes	yes (sort of)	no	yes
intuitionistic analysis	yes	yes	sort of	sort of	yes	no	yes	no	no
smooth infinitesimal analysis	yes	consistent	sort of	sort of	yes	no	unknown	no	yes (sort of)
point-based predicative	yes	yes	yes	yes	no	yes	yes (sort of)	yes (sort of)	yes/no
semi-Aristotelian	yes	yes	no	no	no	yes	yes	yes	yes/no

References

[1] Adžić, M. and M. Arsenijević, "Gunkology and pointilism: Two mutually supervening models of the region-based and point-based theory of the infinite two-dimensional continuum", in Giovanni Macchia, Francesco Orilia, and Vincenzo Fano (eds.), *Space and Time: A Priori and a Posteriori Studies* (Berlin, De Gruyter, 2014), pp. 137–70.

[2] Annas, J., *Aristotle's Metaphysics: Books M and N* (Oxford, Clarendon Press, 1976).

[3] *The Basic Works of Aristotle*, (ed.) R. McKeon (New York, Random House, 1941).

[4] Armstrong, David M., *A World of States of Affairs* (Cambridge, Cambridge University Press, 1997).

[5] Arntzenius, Frank, "Gunk, topology, and measure", Dean W. Zimmerman (ed.), in *Oxford Studies in Metaphysics 4* (Oxford, Oxford University Press, 2008), 225–47.

[6] Arntzenius, Frank, *Space, Time and Stuff* (Oxford, Oxford University Press, 2012).

[7] Arntzenius, Frank and John Hawthorne, "Gunk and continuous variation", *The Monist* 88 (2005): 441–65.

[8] Arsenijević M. and M. Kapetanović, "The 'Great Struggle' between Cantorians and Neo-Aristotelians: Much ado about nothing", *Grazier Philosophische Studien* 76 (2008): 79–90.

[9] Bach, Emmon, "Natural language metaphysics", in R. Barcan Marcus, G. J. W. Dorn, and P. Weingartner (eds.), *Logic, Methodology, and Philosophy of Science VII* (Amsterdam, North Holland, 1986), pp. 573–95.

[10] Baker, Alan, "Simplicity", *Stanford Internet Encyclopedia of Philosophy* (http://plato.stanford.edu/entries/simplicity).

[11] Barker, Chris, "Free choice permission as resource-sensitive reasoning", *Semantics and Pragmatics* 3 (2010), 1–38.

[12] Beeson, Michael J., *Foundations of Constructive Mathematics: Metamathematical Studies* (Berlin, Springer, 1985).

[13] Bell, J. L., *A Primer of Infinitesimal Analysis* (Cambridge, Cambridge University Press, 1998).

[14] Bell, J. L., "The continuum in smooth infinitesimal analysis", in P. Shuster, U. Berger, and H. Osswald, (eds.), *Reuniting the Antipodes: Constructive and Nonstandard Views of the Continuum* (The Hague, Kluwer, 2001), pp. 19–24.

[15] Bell, J. L., "Cohesiveness", *Intellectica* 51 (2009): 145–68.

[16] Benacerraf, Paul and Hilary Putnam (eds.), *Philosophy of Mathematics,* second edition (Cambridge, Cambridge University Press, 1950).

[17] Bennett, B., "A categorical axiomatization of region-based geometry", *Fundamenta Informaticae* 46 (2001): 145–58.

[18] Biacino, L. and G. Gerla, "Connection structures: Grzegorczyk's and Whitehead's definitions of point", *Notre Dame Journal of Formal Logic* 37 (1996): 431–9.

[19] Bishop, E. *Foundations of Constructive Analysis* (New York, McGraw-Hill, 1967).

[20] Bohn, Einar Duenger, "Must there be a top level", *Philosophical Quarterly* 59 (2009), 193–201.

[21] Bohn, Einar Duenger, "An argument against the necessity of unrestricted composition", *Analysis* 69 (2009): 27–31.

[22] Bohn, Einar Duenger, "The necessity of universalism versus the possibility of junky worlds: A rejoinder", *Analysis* 70 (2010): 296–8.

[23] Borgo, Stefano, "Full mereogeometries", *Review of Symbolic Logic* 3 (2010), 521–67.

[24] Bostock, David, "Whitehead and Russell on points", *Philosophia Mathematica* (3)18 (2010): 1–52.

[25] Brentano, Franz, *Philosophical Investigations on Space, Time and the Continuum*, trans. Barry Smith (New York, Routledge, 2010).

[26] Bridges, Douglas and Fred Richman, *Varieties of Constructive Mathematics* (Cambridge, Cambridge University Press, 1987).

[27] Cantor, G., *Grundlagen einer allgemeinen Mannigfaltigkeitslehre. Ein mathematisch-philosophischer Versuch in der Lehre des Unendlichen* (Leipzig, Teubner, 1883).

[28] Cantor, G., "Mitteilugen zur Lehre vom Transfinitten 1, II", *Zeitschrift für Philosophie und philosophische Kritik* 91 (1887), 81–125, 252–70; 92, 250–65; reprinted in [29], 378–439.

[29] Cantor, G., *Gesammelte Abhandlungen mathematischen und philosophischen Inhalts*, ed. E. Zermelo (Berlin, Springer, 1932).

[30] Carnap, Rudolf, "Empiricism, semantics, and ontology", *Revue Internationale de Philosophie* 4 (1950), 20–40; reprinted in [16], pp. 241–57.

[31] Cartwright, R., "Scattered objects", in *Analysis and Metaphysics*, ed. Keith Lehrer (Dordrecht: Reidel, 1975), 153–71.

[32] Chalmers, David, "Verbal disputes", *Philosophical Review* 120 (2011): 515–56.

[33] Chisholm, Roderick M., "Boundaries as dependent particulars", *Grazer Philosophische Studien* 20 (1983): 87–95.

[34] Chisholm, Roderick M., "Boundaries", in *Metaphysics*, by R. M. Chisholm (Minneapolis, University of Minnesota Press, 1989), pp. 83–9.

[35] Corcoran, John, "On definitional equivalence and related topics", *History and Philosophy of Logic* 1 (1980): 231–4.

[36] de Laguna, T., "Point, line, and surface, as sets of solids", *Journal of Philosophy* 19 (1922): 449–61.

[37] Dedekind, R., *Continuity and Irrational Numbers*, translation of *Stetigkeit und irrationale Zahlen* (1901), in *Essays on the Theory of Numbers,* ed. and trans. W. W. Beman (Dover, 1963).

[38] Dorr, Cian, "What we disagree about when we disagree about ontology", in Mark Eli Kalderon (ed.), *Fictionalism in Metaphysics* (Oxford, Oxford University Press, 2005), pp. 234–86.

[39] Dummett, Michael, *Elements of Intuitionism*, second edition (Oxford, Oxford University Press, 2000).

[40] Euclid, *The Thirteen Books of Euclid's Elements*, translation and commentary by Thomas L. Heath, second revised edition (New York, Dover Publications, 1956).

[41] Ewald, William (ed.), *From Kant to Hilbert: A Sourcebook in the Foundations of Mathematics* (Oxford, Oxford University Press, 1996).

[42] Feferman, Solomon and Geoffrey Hellman, "Predicative foundations of arithmetic", *Journal of Philosophical Logic 24* (1995): 1–17.

[43] Feferman, Solomon and Geoffrey Hellman, "Challenges to predicative foundations of arithmetic", in Gila Sher and Richard Tieszen (eds.), *Between Logic and Intuition: Essays in Honor of Charles* (Cambridge, Cambridge University Press, 2000), pp. 317–38.

[44] Fine, Kit, "Our knowledge of mathematical objects", in T. S. Gendler and J. Hawthorne (eds.), *Oxford Studies in Epistemology, Volume 1* (Oxford, Oxford University Press, 2005), pp. 89–110.

[45] Friedman, Michael, *Foundations of Space-Time Theories: Relativistic Physics and Philosophy of Science* (Princeton, Princeton University Press, 1983).

[46] Gardner, R., "Hyperbolic geometry" (http://faculty.etsu.edu/gardnerr/noneuclidean/hyperbolic.pdf).

[47] Gauss, Carl Friedrich, *Briefwechsel mit Schumacher* (Leipzig, Teubner, 1899. [New York, Johnson Reprint Corp., 1972]), Bd. II, p. 268.

[48] Gierz, G. et al., *A Compendium of Continuous Lattices* (Berlin, Springer, 1980).

[49] Gruszczyński, R. and A. Pietruszczak, "Full development of Tarski's geometry of solids", *Bulletin of Symbolic Logic* 14 (2008): 481–540.

[50] Gruszczyński, R. and A. Pietruszczak, "Space, points, and mereology: On foundations of point-free Euclidean geometry", *Logic and Logical Philosophy* 18 (2009): 145–88.

[51] Grzegorczyk, A., "Axiomatizability of geometry without points", *Synthese* 12 (1960): 228–35.

[52] Hawthorne, John, "Before-effect and Zeno causality", *Noûs* 34 (2000): 622–33.

[53] Heck, Richard, "Cardinality, counting, and equinumerosity", *Notre Dame Journal of Formal Logic* 41 (2000): 187—209; reprinted in [54], pp. 156–79.

[54] Heck, Richard, *Frege's Theorem* (Oxford, Oxford University Press, 2011).

[55] Hellman, Geoffrey, "Never say 'never'!: On the communication problem between intuitionism and classicism", *Philosophical Topics* 17 (1989): 47–67.

[56] Hellman, G., *Mathematics without Numbers: Towards a Modal-Structural Interpretation* (Oxford, Oxford University Press, 1989).

[57] Hellman, G., "Structuralism without structures", *Philosophia Mathematica* (3)4 (1996): 100–23.

[58] Hellman, G., "Mathematical pluralism: The case of smooth infinitesimal analysis", *Journal of Philosophical Logic* 35 (2006): 621–51.

[59] Hellman, Geoffrey and Stewart Shapiro, "Towards a point-free account of the continuous", *Iyyun: The Jerusalem Philosophical Quarterly* 61 (2012): 263–87.

[60] Hellman, Geoffrey and Stewart Shapiro, "The classical continuum without points", *Review of Symbolic Logic* 6 (2013): 488–512.

[61] Hilbert, David, *Grundlagen der Geometrie* (Leipzig, Teubner); translated by E. Townsend as *Foundations of Geometry* (La Salle, IL, Open Court, 1959).

[62] Hintikka Jaakko, "Potential infinity", *Philosophical Review* 75 (1966): 197–218.

[63] Hirsch, Eli, "Physical-object ontology, verbal disputes, and common sense", *Philosophy and Phenomenological Research* 70 (2005): 67–97.

[64] Hudson, Hud, "Touching", *Philosophical Perspectives Metaphysics* 15 (2001): 119–28.

[65] Hudson, Hud, *The Metaphysics of Hyperspace* (Oxford, Oxford University Press, 2005).

[66] Jané, Ignacio, "Idealist and realist elements in Cantor's approach to set theory", *Philosophia Mathematica* (3)18 (2010): 193–226.

[67] Johnstone, P. T., "The point of pointless topology", *Bulletin (New Series) of the American Mathematical Society* 8(1) (1983): 41–53.

[68] Knorr, Wilbur R., "Infinity and continuity: The interaction of mathematics and philosophy in antiquity", in Normal Kretzmann (ed.), *Infinity and Continuity in Ancient and Medieval Thought* (Ithaca, Cornell University Press, 1982), pp. 112–45.

[69] Kripke, Saul, "Identity and necessity", in Milton K. Munitz (ed.), *Identity and Individuation* (New York, New York University Press), pp. 135–64.

[70] Lear, Jonathan, "Aristotelian infinity", *Proceedings of the Aristotelian Society, new series* 80 (1980), 187–210.

[71] Lear, Jonathan, "Aristotle's philosophy of mathematics", *Philosophical Review* 41 (1982), 161–92.

[72] Leibniz, Gottfried, *Mathematische Scriften von Gottfried Wilhelm Leibniz*, ed. C. I. Gerhart (Berlin, A. Asher, H. Halle, W. Schmidt, 1863).

[73] Leibniz, Gottfried, *Samtliche Schriften und Briefe: Philosophishce Schriften*, Series 6, Volumes 1–3 (Berlin, Akademie-Verlag, 1923–80).

[74] Leibniz, Gottfried, *New Essays on Human Understanding*, ed. and trans. P. Remnant and J. Bennett (New York, Cambridge University Press, 1996).

[75] Leonard, Henry S. and Nelson Goodman, "The calculus of individuals and its uses", *Journal of Symbolic Logic* 5 (1940): 45–55.

[76] Levey, Samuel, "Leibniz on mathematics and the actually infinite division of matter", *Philosophical Review* 107 (1998): 49–96.

[77] Lewis, David, *On the Plurality of Worlds* (Oxford, Blackwell, 1986).

[78] Lewis, D., *Parts of Classes* (Oxford, Blackwell, 1991).

[79] Linnebo, Øystein, "The potential hierarchy of sets", *Review of Symbolic Logic* 6 (2013): 205–28.

[80] Linnebo, Øystein and David Nicolas, "Superplurals in English", *Analysis* 68 (2008): 186–97.

[81] Linnebo, Øystein and Shapiro, Stewart, "Actual and Potential Infinity", *Noûs*, forthcoming.

[82] Maddy, Penelope, *Realism in Mathematics* (Oxford, Oxford University Press, 1990).

[83] Maddy, Penelope, *Second Philosophy: A Naturalistic Method* (Oxford, Oxford University Press, 2007).

[84] Markosian, Ned, "Simples", *Australasian Journal of Philosophy* 76 (1998): 213–28.

[85] Menger, K., "Topology without points", *Rice Institute Pamphlets* 27 (1940): 80–107.

[86] Merricks, Trenton, *Objects and Persons* (Oxford, Oxford University Press, 2001).

[87] Miller, Fred D., "Aristotle against the atomists", in Normal Kretzmann (ed.), *Infinity and Continuity in Ancient and Medieval Thought* (Ithaca Cornell University Press, 1982), pp. 87–111.

[88] Nagel, Ernest, "The meaning of reduction in the natural sciences", in A. Danto and S. Morgenbesser (eds.), *Philosophy of Science* (Cleveland: Meridian Books, 1960), 288–312.

[89] Nagel, Ernest, *The Structure of Science* (New York: Harcourt, Brace, and World, 1961).

[90] O'Neill, Barrett, *Elementary Differential Geometry*, revised second edition (London, Academic Press, 2006).

[91] Pambuccian, Victor, "Sphere tangency as single primitive notion for hyperbolic and Euclidean geometry", *Forum Mathematicum 15* (2003), 943–7.

[92] Parsons, Josh, "Theories of location", *Oxford Studies in Metaphysics 3*, ed. Dean W. Zimmerman (Oxford, Oxford University Press, 2007), pp. 201–32.

[93] Pieri, M., "La geometria elementare instituita sulle nozione di 'punto' e 'sefra'", *Memorie di Matematica e di Fisica della Società Italiana delle Scienze, Serie Terza 15* (1908), 345–450.

[94] Plato, *The Republic of Plato*, translated by Francis Cornford (Oxford, Oxford University Press, 1945).

[95] Proclus, *Commentary on Euclid's elements I*, trans. G. Morrow (Princeton, Princeton University Press, 1970).

[96] Putnam, Hilary, "Mathematics without foundations", *Journal of Philosophy* 64 (1967): 5–22.

[97] Rayo, Agustín, "Beyond plurals", in Agustín Rayo and Gabriel Uzquiano (eds.), *Absolute Generality* (Oxford, Oxford University Press, 2006): 220–54.

[98] Roeper, P., "Region-Based Topology", *Journal of Philosophical Logic* 26 (1997): 251–309.

[99] Roeper, P., "The Aristotelian continuum. A formal characterization", *Notre Dame Journal of Formal Logic* 47 (2006): 211–31.

[100] Rogers, Hartley, *Theory of Recursive Functions and Effective Computability* (New York, McGraw-Hill, 1967).

[101] Russell, Jeffrey Sanford, "The structure of gunk: Adventures in the ontology of space", in Dean W. Zimmerman (ed.), *Oxford Studies in Metaphysics 4* (Oxford, Oxford University Press, 2008), pp. 248–74.

[102] Schaffer, Jonathan, "Monism: The priority of the whole", *Philosophical Review 119* (2010): 31–76.

[103] Scott, Dana, "Extending the topological interpretation to intuitionistic analysis I", *Compositio Mathematica* 20 (1968): 194–210.

[104] Scott, Dana, "Extending the topological interpretation to intuitionistic analysis II", in A. Kino, J. Myhill, and R. Vesley (eds.), *Intuitionism and Proof Theory*, (Amsterdam, North Holland, 1970), pp. 235–55.

[105] Shapiro, Stewart, *Foundations without Foundationalism: A Case for Second-Order Logic* (Oxford, Oxford University Press, 1991).

[106] Shapiro, Stewart, *Philosophy of Mathematics: Structure and Ontology* (New York, Oxford University Press, 1997).

[107] Shapiro, Stewart, *Thinking about Mathematics: The Philosophy of Mathematics* (Oxford, Oxford University Press, 2000).

[108] Shapiro, Stewart, "We hold these truths to be self evident: But what do we mean by that?", *Review of Symbolic Logic* 2 (2009): 175–207.

[109] Shapiro, Stewart, "An 'i' for an i: Singular terms, uniqueness, and reference", *Review of Symbolic Logic* 5 (2012): 380–415.

[110] Shapiro, Stewart, *Varieties of Logic* (Oxford, Oxford University Press, 2014).

[111] Shapiro, Stewart and Geoffrey Hellman, "Frege meets Aristotle: Points as abstracts", *Philosophia Mathematica* (3), 25 (2017): 73–90.

[112] Sider, Theodore, *Four-Dimensionalism* (Oxford, Oxford University Press, 2001).

[113] Sider, Theodore, *Writing the Book of the World* (Oxford, Oxford University Press, 2012).

[114] Simpson, Stephen G., *Subsystems of Second Order Arithmetic: Perspectives in Logic*, second edition (Cambridge, Cambridge University Press, 2009).

[115] Smith, Barry, "Boundaries", in L. Hahn (ed.), *The Philosophy of Roderick M.* (La Salle, Open Court, 1996), pp. 533–61.

[116] Sorabji, Richard, *Time, Creation and the Continuum: Theories in Antiquity and the Early Middle Ages* (Chicago, University of Chicago Press, 2006).

[117] Strawson, P. F., *Individuals: An Essay in Descriptive Metaphysics* (London and New York, Routledge, 1959).

[118] Suárez, Francisco, *Disputationes Metaphysicae, in quibus et universa naturalis theologia ordinate traditur, et quaestiones omnes ad duodecim Aristotelis libros pertinentes accurate disputantur* (Salamanca, Spain, 1597; reprinted by Georg Olms, Hildesheim, 2009).

[119] Tarski, A., "Foundations of the geometry of solids", in *Logic, Semantics, and Metamathematics: Papers from 1923 to 1938* (Oxford, Oxford University Press, 1956).

[120] Troelstra, A. S. and D. van Dalen, *Constructivism in Mathematics*, two volumes (Amsterdam, North Holland Press, 1988).

[121] Uzquiano, Gabriel, "Receptacles", *Philosophical Perspectives* 20 (2006): 427–51.

[122] Uzquiano, G., "Plural quantification and modality", *Proceedings of the Aristotelian Society* 111 (2011): 219–50.

[123] Uzquiano, Gabriel, "Mereological harmony", in Karen Bennett and Dean W. Zimmerman (eds.), *Oxford Studies in Metaphysics 6* (Oxford, Oxford University Press, 2011), pp. 199–224.

[124] Van Inwagen, Peter, *Material Beings* (Ithaca, Cornell University Press, 1990).

[125] Varzi, Achille, "Boundaries, Continuity, and Contact", *Noûs* 31 (1997): 26–58.

[126] Visser, Albert, "Interpretability degrees of finitely axiomatized sequential theories", *Archive for Mathematical Logic* 53 (1–2) (2014): 23–42.

[127] White, Michael J., *The Continuous and the Discrete: Ancient Physical Theories from a Contemporary Perspective* (Oxford, Oxford University Press, 1992).

[128] Whitehead, Alfred North, *The Organisation of Thought: Educational and Scientific* (London, Williams and Norgate, 1917).

[129] Whitehead, Alfred North, *An Enquiry Concerning the Principles of Natural Knowledge* (Cambridge: Cambridge University Press, 1919).

[130] Whitehead, Alfred North, *The Aims of Education and Other Essays* (New York, Macmillan, 1929).

[131] Whitehead, Alfred North, *Process and Reality* (New York Macmillan Company, 1929).

[132] Williamson, Timothy, *Modal Logic as Metaphysics* (Oxford, Oxford University Press, 2013).

[133] Wilson, Mark, *Wandering Significance* (Oxford, Oxford University Press, 2006).

[134] Zimmerman, Dean, "Could extended objects be made out of simple parts? An argument for 'atomless gunk'", *Philosophy and Phenomenological Research* 56 (1996): 1–29.

[135] Zimmerman, Dean, "Indivisible parts and extended objects: Some philosophical episodes from topology's prehistory". *The Monist* 79 (1996): 148–80.

Index